Electric Dreams

Published by 404 Ink Limited
www.404Ink.com
@404Ink

Please note: Some references include URLs which may change or be unavailable
after publication of this book. All references within endnotes were accessible and
accurate as of January 2024 but may experience link rot from there on in.

Editing: Heather McDaid
Typesetting: Laura Jones-Rivera
Proofreading: Laura Jones-Rivera
Cover design: Luke Bird
Co-founders and publishers of 404 Ink:
Heather McDaid & Laura Jones-Rivera

Print ISBN: 978-1-912489-86-2
Ebook ISBN: 978-1-912489-87-9

Printed and bound in Great Britain by Clays Ltd, Elcograf S.p.A.

404 Ink acknowledges and is thankful for support from
Creative Scotland in the publication of this title.

LOTTERY FUNDED

Electric Dreams

*Sex Robots and the Failed
Promises of Capitalism*

Heather Parry

Inklings

We can be responsible for machines;
they do not dominate or threaten us.
We are responsible for boundaries; we are they.

– Donna Haraway, *The Cyborg Manifesto*

Contents

Introduction

Within mere decades, it will be totally normal for human beings to be in romantic and sexual relationships with robots. This is the contention of David Levy, chess master, sometime businessman and author of a book lauded by *the Guardian*, *the Telegraph* and *the Washington Post*—a book I read several years ago, and became quietly obsessed with. 2008's *Love and Sex with Robots* started life as Levy's PhD thesis and made grand claims that might sit perfectly alongside the wildest statements from the Elon Musks of this world but seem otherwise so deluded that my copy is decorated with about a hundred astonished Post-Its, scrawled with things like *where is the evidence for any of this?* and, more than once, *unhinged!* The claim is that by 2050, 'Humans will fall in love with robots, humans will marry robots, and humans will have sex with robots, all as (what will be regarded as) 'normal' extensions of our feelings of love and sexual desire for other humans.'[1]

This is not a fringe view—or at least, it isn't treated as such. Countless articles have been written about the advent of AI-infused sex robots, and documentaries are made about the small circle of men spearheading their creation—or, rather, spearheading the manufacture of impressively rendered silicon sex dolls with basic robotic heads, with pre-loaded personalities to chose from, which is about as close as we've got so far. The eventual existence of sex robots that are 'indistinguishable from humans' is treated as a serious intellectual consideration, and, more than this, a threat; the Campaign Against Sex Robots, as an example, demands the 'end of sex robots' before they have even really begun to emerge. All seem to agree, as Levy claims, that 'love and sex with robots on a grand scale are inevitable.'[2]

Sex robots do, however, already exist in one form: as an irresistible concept, an idea that we buy into—as a society—to various degrees time and time again. We accept them as an certainty, a tantalizing prospect. This is worth investigation, because what do sex robots actually promise us? A regurgitation of regressive ideas about what women are, what men are, what pleasure is. They appeal to a viewpoint that believes that ownership of bodies will bring satisfaction, but only to a thin and unsatisfying idea of what sexuality looks like or can be. Even when we oppose them, we rely on arguments that demean others, on biological essentialist talking points that negatively affect the discussion of women's rights, and, on every level, we buy into the ludicrous dreams of tech companies and entrepreneurs.

There are a variety of social systems holding up the sex robot as a future promise, and as the contract of capitalism breaks down—as inequality spirals out of control, climate breakdown accelerates, and capitalist economic ideology fails to deliver its promised riches to the masses—we cling to sex robots as the (il)logical end point of systems we know, even as those systems wither out of relevance. The promises they make are unfeasible ones. To reject these claims is easy; what's more interesting is to look at what the existence of this obsession says about our society today. If we look at sex robots from non-capitalist or non-heteronormative lenses— from queer, anti-colonial, collectivist perspectives—we can begin to see how they attempt to entrench dying systems, prompt political regression, and stop us from considering what else we might be working towards. If we let these bizarre fantasies fade into the ether, we can discover, instead, what the future can really hold, and how we might truly come together. No pun intended.

Chapter 1
Sex robots as science fiction

In the late 1990s, real robots and their human creators burst into the British public's consciousness, thanks to a show filmed in a Docklands warehouse on a budget of about three pounds. *Robot Wars*, BBC 2's cult classic 'robot combat competition' pitted amateur robotics engineers—or rather, their remote-controlled inventions—against each other in battle. The theatre of conflict was a flame-laden, obstacle-heavy arena policed by the 'House Robots'; professionally-designed machines with names like Sir Killalot, Mr. Psycho and Sergeant Bash. Watching the footage now, it's impossible not to develop extreme fondness for the amateur roboticists, usually men between 18 and 50 with names like Pete and Steve and Alan, sometimes with their partners, sometimes, their kids. The contestants have little to no stage presence and take their hobby *extremely* seriously; very often Craig Charles, a presenter of enormous charisma, has a hard time pulling more than single word answers out of them. But they are highly affable and thrilled to be there, and all of them have spent months designing and building robots with names like Behemoth and Napalm and Angel of Death, only to smash them to bits or see them set on fire within a matter of minutes, sometimes seconds.

There are precious few 'walkerbots'—robots with legs—on any series. In fact, the most indomitable robots were ones like Razer, a wedge with a crushing scorpion's tail; Chaos 2, a dual wedge with a powerful pneumatic flipping arm; and Hypno-Disc, an armoured rectangle with a wildly destructive horizontal flywheel on the front. Even the House Robots, built by professionals, relied on speed, weight, crushing ability or some sort of weapon; pincers, hammers or pneumatic tusks. They were hefty.

Not a single one of the robots on *Robot Wars* had anything humanoid about them; twenty-five years later, the same is true of the robots we now live and work with on a regular basis. Robots are everywhere; they are in factories, shifting heavy pallets; they are huge mechanical arms with, at most, six degrees of articulation, painting cars; they are featureless disks hoovering your home. As the University of Edinburgh's Professor Adam Stokes puts it, 'robots are good at dangerous, dull and dirty jobs.'[3] And precisely none of these jobs currently require them to look human.

It's not just that humanoid appearance isn't necessary; it's that the human shape, broadly speaking, is completely inefficient. The amateur engineers of *Robot Wars* knew it, and those working in the field decades later know it too. Humans can be easily knocked over; we often trip and fall. We require an enormous amount of energy to do anything. All of our vital, fleshy, vulnerable major organs are right in our highly stabbable torsos, with only a few layers of skin, fat, muscle and sometimes bone between them and a deadly implement. The mechanism of the human body isn't a sensible one; as Simon Watson, Lecturer in Robotics at the University of Manchester puts it,

> We like to think that we're the dominant creatures on the planet, so mobile robots should look like us. But the fact is, they shouldn't. We can't fly, we're not very good swimmers, we can't live in a vacuum and if we want to travel more than a mile, most of us will get on some type of wheeled vehicles. Bipedal locomotion has served us well but it is limited and requires a huge amount of brain power and years of learning to perfect…. After nearly 100 years of development, our most advanced humanoid robots can only just open a door without falling over.[4]

The last sentence here is not an exaggeration; in 2022 the news of a robot opening a door was splashed across media outlets with great fanfare. Despite our inefficiency, vulnerability and illogical design, the human body is a wildly, nonsensically capable object. When you break down what is required of a machine to reproduce even the most basic human tasks—like getting up off the floor, picking something off a high shelf, or, indeed, opening a door—you start to realise why people believe in a God. Not to be arrogant, but it is simply mind-boggling what evolution has managed to create. We float about the world doing our jobs and hugging our loved ones and operating heavy machinery with barely a thought given to the complex, intricate and numerous systems at work. Take the human hand. Our hands have twenty-six degrees of freedom; to recreate this mechanically, every one degree would require individual control. We do it without even noticing. Neuroscientists estimate that humans have between twenty-one and thirty-three distinct senses, many working in tandem to ensure we can move about the world as we currently do. Without even one or two of them, we could be completely incapacitated in ways we can barely comprehend.

A favourite example is that of proprioception, which allows us to instantaneously sense and understand our body's movement, location and actions. This might not sound that momentous, but that's because we can't really conceive of not having that sense; indeed, there are only a handful of people in the world who don't have any proprioception. The most famous is probably Ian Waterman, who lost this sense at 19, when a viral illness caused the production of an antibody which destroyed the nerves that told his brain what his body was doing. Though there was nothing wrong with the physical mechanisms of his limbs, he woke up after three days apparently unable to move. It was discovered that he actually *could* move his body, but not coordinate its movements; he had no feedback as to what his limbs' locations or actions were. Ian learned to sit up, feed himself and, amazingly, walk again after laboriously retraining his mind and body, relying totally on visual feedback to know where his limbs were and what they were doing. Since losing his proprioception in 1971, he has lived a life of painstaking choreography, and still his stability is fragile; if an object is heavier than anticipated, it throws off his balance completely. If the lights go off in a room while he's standing, he will collapse.

Proprioception is just one sense that humanoid robots would have to recreate to move about the world. Vision is another, and though this might seem to be as 'simple' as installing a camera, the reality is very different. It takes the human brain just 13 milliseconds to process an image captured by the eye; a five-megapixel camera will give a robot 5 million numbers to interpret per single image, with the camera sending between twenty and fifty images per second. All this data needs to be processed and stored. This is before we consider the requirements of recognising anything within those images.

It is not just the complexity of building humanoid robots that works against their potential existence; it is the running of them too. As anyone with an iPhone will know, our ability to create exciting tech far outstrips our ability to make batteries that can power it for any decent amount of time. The processing of one hundred million numbers per second is incredibly energy intensive, and that is just for one second of vision for a humanoid robot. When you add in the energy requirements of bipedal motion, we are talking about truly enormous amounts of power; even the most efficient biped robots are three times less efficient than humans at converting energy to movement. In addition, there is simply so much going on inside robots that needs to be powered; it's estimated that over fifty percent of a moving robot's power goes towards its myriad sensors, controllers and computers. Spot, the 'digidog' from Boston Dynamics—the much-hyped yellow quadroped robot often seen scuttling over terrain and being assaulted by hockey sticks in the company's promotional videos—has a best run time of only 90 mins. This didn't stop the NYPD from buying two for $750,000 each in 2023, but it'll be interesting to see how that potential run time translates to real world usage. The company's humanoid demonstrator Atlas weighs as much as Spot; though they don't specify, we can estimate its run time at likely 30 mins or less. When you compare this to how easily a person can power themselves for sixteen hours on about 2000 calories, all of which can be gained via eating a relatively small amount of food (supplemented by a couple of cups of coffee), you do sort of ask yourself what the hell we're doing trying to recreate something even vaguely similar when *we* already exist, and when our existence can only be as a result of some grand cosmic miracle that we do not yet fully understand.

The resounding message from robotics scholars is that it makes no sense to make a robot like a human, and even if we did want to do it, we can't. As Professor Ruth Aylett told an audience at Heriot-Watt University's Robotarium, on the topic of making human-like robots,

> Let me let you into a secret, not much of a secret amongst people who do robotics, I will tell you: We cannot do that. We have no prospect of doing that in the immediate future. The more you do robotics, the more wonderful you see living things are, even quite simple ones.[5]

This is not an unusual claim amongst roboticists and computer scientists: currently, the concept of a functional human-like robot is nothing but science fiction. Despite this, the common narrative of public robotics discourse is that humanoid robots are imminent. There are countless movies, books and other fictions written about the seemingly inevitable point at which robots will be indistinguishable from humans, which the media leads us to believe is very much on the horizon, if not already here. Social media compounds this assumption; as I write, a viral video claims to show Hamad bin Isa Al Klalifa, the King of Bahrain, being escorted by his enormous and highly militarised robot bodyguard. 'It can speak 6 languages,' claim the many versions of the text alongside this video. 'It's armed with an electric teaser, an undisclosed 360° camera system, 3 hidden machine guns & ammunition to fight 1050 men. Cost $7.4 million.' The truth is that the 'robot', Titan, is in fact an actor in a costume that you can hire for corporate events, and the man walking in front of him is not the King of Bahrain, but Mohammed bin Rashid Al Maktoum, the 'Sheik' of Dubai.[6] Aside from the obvious and troubling racial politics at play here, this particular incident highlights exactly how easy it is to convince people that functional humanoid robots fit for daily use are already here. On some level, we all seem to believe it.

* * *

The media hype around robots is almost entirely reserved for the (presumed) existence, forthcoming or current, of humanoid machines. It's difficult to imagine a video of a Roomba, or robot from a car factory, or an industrial robot going viral today; the novelty has long since worn off. As Professor Adam Stokes at the Edinburgh Centre for Robotics explains:

> What we expect robots to do is fed from a basis of science fiction. Robots as agents are around us all the time. Robots as embodied systems have a range of uses but they're not that interesting to the general public.[7]

One notable exception to this is a robot artwork by controversy-courting Chinese artists Sun Yuan and Peng Yu, entitled *Can't Help Myself (2016-19)*, which was commissioned for installation at the Guggenheim Museum. An industrial robot arm was positioned in the middle of a glass box, with a puddle of deep red liquid around it. The robot's task was to keep the liquid within a pre-programmed circular area around it; if the liquid seeped out of this area, the robotic arm turned and swept the liquid back towards the middle of the circle. The process of this Sisyphean task eventually splattered the insides of the glass box with the blood-like liquid, evoking a scene of real horror. The artists intended this to be a critique of the

increasing role of technology in closing borders and crim-
inalising migrants; as curator and writer Xiaoyu Weng
writes for the official Guggenheim description of the
artwork, 'bloodstain-like marks that accumulate around
[the robotic arm] evoke the violence that results from
surveilling and guarding border zone'.[8] This intended
meaning was completely drowned out when a video of the
machine went viral after its appearance at the 2019 Venice
Biennale. In the years since its creation, the robotic arm
had slowed, and the inside of the glass box was stained
by three years' worth of the vicious red liquid. Without
the context of the show, social media users immediately
and intensely personified the robot, claiming that it was
leaking the red liquid and desperately trying to contain it
to keep itself running, and that it had once been free to
dance, but now was too tired, too worn down to do so.
Many claimed they had never been so moved by a work
of art. To some extent, the artists did court this anthro-
pomorphism; there's something Xenomorphic about the
design of the robot, and it was programmed with movements known as 'ass shake' and 'scratch an itch'. But no
matter how it moves, the artwork is simply an industrial
robotics arm, and is not intended to convey emotion; the
whole point is that it is an inhuman thing, performing an
inhuman act. Still, we project onto it.

What this shows is that, no matter how nakedly
mechanical a robot, nor how un-humanlike it looks,

we simply cannot get away from that idea that robots have—or perhaps will have—*some* sort of humanity. In fact, the first piece of popular media to address the concept of artificial, human-created people (which introduced the term *robot* to the English language) also popularised the idea that they would, in some way, be just like us. Czech writer Karel Čapek's 1920 play *Rossum's Universal Robots* told the story of *roboti*, a race of workers created from synthetic material, who eventually rise up against their human oppressors and cause their destruction. This might have been the first story of machines developing human-like mental abilities and emotions—in this instance, rage at the injustice of their treatment, and an ability to organise against those who do them harm—but it would certainly not be the last. Myriad books, films, plays and artworks on the topic of robots have been created in the century since then, and in the vast majority of them, the robots develop humanistic consciousness.

This is a quite logical extension of how we, as humans, deal with the problem of other people generally. As beings trapped in our own subjective experiences of the world, we can never actually know that any other people have minds in the same way that we have minds, or indeed that these other people have minds at all. Logically speaking, all human-looking creatures around us at any time could be automatons. Their interior mental

15

states are, necessarily, completely inaccessible to us. But it's impossible to live under that logic (at least without being a total prick), so instead we engage in a type of inductive inference: *I know that I move around the world in a human body and I know that I have thoughts, feelings, emotions; that I have a mind. Therefore, I must infer that the other people moving around the world in human bodies, exhibiting the behaviours that imply they have an internal mental world, also have minds.* We have trained ourselves, since infancy, to interact with others on the assumption that they have minds comparable to our own, and so have the same capacity for anger, sadness, joy, and love as we do. So, what happens when we are presented with the idea (and it is just an idea) of a robot that looks and moves and walks and talks just like us? We assume, even unconsciously, that they will have internal worlds comparable to ours. When we make this connection, we then imagine them having relationships with us too, whether adversarial, friendly, romantic—or sexual.

It's not difficult to recall stories that feature robots as sexualised beings. Only six years after Čapek's play, the Fritz Lang movie *Metropolis* showed the Maria robot as emotionless and sexualised, eventually embracing her inventor Rotwang. 1972's *The Stepford Wives* by Ira Levin centres on a fleet of robot women created solely to please their suburban American husbands. Philip K Dick's 1968 novel *Do Androids Dream of Electric Sheep?*—later brought to screen as *Blade Runner*—features android bounty hunter Deckard sleeping with android Rachael. Released in 1993, Isaac Asmiov's *Forward the Foundation* features a historian with a robot wife named Dors. Since the Millennium, there have been increasing numbers of these stories, with people, most often men, falling in love with mostly female-coded robots or artificial intelligences; *Her, Lars at the Real Girl* and *Ex Machina* are just three examples. The discourse around these stories always seems to be about the intricacies of these relationships—what does consent look like when you're dating a non-person? How do you have sex with an android? What is the feminist perspective on female coded chatbots who are designed to pander to male egos?—rather than the fundamental question of whether human-robot relationships would ever occur. If they are going to exist, we seem to be saying, and if they are going to have minds like ours, then *of course* we'll fuck them.

Perhaps all of this explains why a book like *Love and Sex with Robots*—with its highly questionable central premise—can not only be published, but be lauded by the media establishment. Science fiction has always liked to make grand claims about when robots will become ubiquitous; even *Rossum's Universal Robots* was set in the year 2000, when, it said, robots would be cheap and everywhere around us. *Love and Sex with Robots,* does the same: remember, by 2050, Levy insists that we will be loving, marrying and fucking robots, and this will be perfectly normal. As I write, there are as many years between now and 2050 as there are between this moment and the first series of *Robot Wars*. It is only nine years until Dr Levy thinks an android might be crowned *Time's* Person of the Year. In spring 2023, UCLA made a grand fuss of its robot ARTEMIS who can walk, allegedly, at a speed of 2.1 metres per second, and withstand challenges from human bodies. But their promotional videos show a creation that looks not dissimilar to a Tesla-designed version of Johnny Five from problematic 1986 movie *Short Circuit*, and though, yes, it has legs and can walk, it doesn't look in the least bit humanoid. For all the fanfare, ARTEMIS looks like your grandmother could take him out with a swift kick to the knees, and probably outrun him too.

The idea that we might realistically have humanoid robots around by 2050 is sheer fantasy, as roboticists themselves are telling us, and the claim that they'll be so advanced we'll be marrying them is surely delusional. There is no real reason to make a humanoid robot, and if we wanted to, we couldn't, and if we did, we couldn't power it, and if we do manage to get over any of these significant hurdles, let alone all of them, then that moment won't come any time soon. This is what the experts are telling us. Yet *Love and Sex with Robots* and its claims were reviewed completely credulously. Technology journalist Julian Dibbell said in *The Telegraph*: 'It's no mean feat just presenting a prediction as outlandish as that as unabashedly as Levy does. But more impressive still is how coherently he backs it up.'[9] *The Chicago Sun-Times* said: 'The deeper you get into the book, the more difficult it becomes to dismiss his thesis.'[10] No matter what the facts, it seems, we—and by *we* I mean both the public at large and the intellectual establishment—simply love a fiction.

By now, you might be somewhat confused. Here we are saying that humanoid robots, and by extension sex robots, are currently impossible, and in the near future unfeasible. And yet, you might be thinking, can't you already buy them?

The simple answer is: no, you cannot buy a sex robot. You *can* buy a 'fucking machine', such as the F Machine Pro 3, which can be yours for under £600 and which, according to its Amazon page, got a 'massive superfuck upgrade' from the previous model, the Pro II. These machines, which are surely more deserving of the term 'sex robot' than anything else currently on the market, take the form of a hydraulic arm onto which you attach a dildo. Position the machine accordingly and… well, off you go. But I don't think it's the F Machine Pro 3 which David Levy imagines us walking down the aisle with. What the many articles and books on this subject mean when they say 'sex robot' is, most of the time, Harmony.

Harmony is billed as 'the world's first sex robot', and she's the product of Abyss Creations under CEO Matt McMullen, the Elon Musk of the fuckable tech world. Despite Harmony's billing, she's anything but a robot; in fact, she's a RealDoll (a hyper-realistic life-size posable mannequin with silicone flesh) with an animatronic head. The combination of the doll and a RealBotix head, which is manufactured by RealDoll in association with two other companies, is known as the RealDollX; it is designed to run with 'customisable AI software' controlled via an app, letting you make decisions about the head's personality and voice. Harmony as she is presented in the media has been given a Scottish accent—specifically, according to the company, a 'Glasgow uni' accent. Make of that what you will.

There is a range of male dolls on the RealDoll website, but there are far fewer options than for the female line, and the quality is almost comically bad; all resemble the Tracy brothers from *Thunderbirds*. There are no male RealBotix heads available at the time of writing; it's even less possible to claim there are male sex robots than it is to claim there are female ones. Harmony, without any customisation, will currently set you back about $6,000. If you want to enlarge her boobs, give her pubic hair, and give her five more of the 'SenseX' inserts that will cause her animatronic face to respond to your touch, you're closer to $10,000. The app is $2.50 a month, and for

$50 a month you can take out insurance which will cover '$2,000 of damage, up to two incidents a year.'

It's something of a bewildering experience to watch scientists, presenters, salespeople and journalists speak of the incredible achievement of the RealBotix 'robotic heads' while also watching them in action. The heads do have ten degrees of freedom (eyes, mouth, neck and eyebrow movements), but this is nothing compared to the forty-six mimic muscles on the human face, and this extremely limited range of expressive movement makes the RealBotix heads seem slack and uncanny, like murderers wearing the shorn-off faces of their victims. Trying to match the wild claims about their engineering sophistication to the actual experience of seeing them move is an exercise in losing your mind. You become grateful that from the neck down, these dolls are nothing more than unmoving silicone on PVC skeleton frames.

And yet, they are still the closest things we have to the sex robots we are, endlessly, promised. We have only a couple of decades for these robot-headed fuck dolls to be turned into the 'highly sophisticated humanoid robots' we have been promised.[11] By 2050, the oft-cited futurist, best-selling writer and Silicon Valley favourite Ray Kurzweil reckons that machine intelligence will already have vastly outstripped human intelligence in such a way that human civilisation will be irreparably changed, in a moment known popularly as 'the Singularity'. We'll be able to create whole virtual bodies with nanobots, says Kurzweil, and we'll have already had twenty years of putting nanobots inside our bodies, to 'basically wipe out disease'.[12] We will already have nanotech foglets that can make food out of thin air (predicted for the 2040s), be able to upload our minds, our entire consciousness, to the cloud (predicted for the 2030s) and we will have glasses which beam images directly onto the retina (which was supposed to happen in the 2010s).

These are bold claims, but they're not only held by the people who write them. These claims get reported on, repeated and quoted as if they are inarguable. We treat them as if they are just around the corner. If it's hard to imagine believing this, consider another set of predictions for the same time period. According to NASA, by 2050 we will have warmed the planet by 1.5 degrees Celsius. Sea level will have risen by thirty centimetres, swamping coastal regions.[13] 70-90% of our coral reefs will be dead. Entire ecosystems will have collapsed, crops will fail, parts of the planet will become uninhabitable and tens of millions of people will be displaced. The most recent reports, in fact, suggest that we might reach this point by the late 2020s.[14] It's a lot easier on the mind and body to believe that we'll be uploading our brains to the network, ridding ourselves of disease and marrying super-hot android robots that will fulfil every sexual desire we have ever had, isn't it? A lot nicer. Fiction is, after all, how we escape reality.

Chapter 2

Sex robots as symbols of unsatisfied heterosexuality

Who really wants to fuck a robot? If you propose this question at the pub, you'll get a lot more expressions of interest than anticipated; it's a novel idea, and who hasn't had a sexual experience or two just for the sheer novelty? But take apart what it would actually mean to fuck a robot (putting your very sensitive genitalia in, on or near some heavy machinery), and the interest diminishes; point out that you can actually fuck a robot for less than £600, but it will be the penetrating party, and suddenly a lot of people aren't so keen. Our interest in robot sex is seemingly dependent on a very narrow set of criteria, and these criteria are largely heteronormative.

It's not an overreach to say that sex robots, or the idea of them, is a primarily (but not exclusively) male interest.

Abyss Creations, the producers of the RealDollX, estimates that over 95% of their customers are men, even considering the small line of male dolls that they offer—around 10% of their current output.[15] (You can buy trans dolls from Abyss Creations, though you have to custom order these.) There's little data on whether it is women or men buying these few male dolls (and none addressing potential trans customers), but the way they are posed on the RealDoll website—bent over, underwear pulled teasingly down; holding their erections in a way that apes promotional screenshots from guy-on-guy porn—suggests at least a small gay male clientele. This isn't really a surprise, given that, as computer scientist and sex robot expert Kate Devlin writes, 'they are built to serve the male gaze'; some of that, of course, includes the male gays.[16] CEO Matt McCullen claims they're working on male robots able to be programmed as gay or straight, as well as the first lesbian version of their female sex dolls, but given how Abyss creations views heterosexual desire, it's hard to imagine their understanding of queer desire would be anything but abysmal. The company seems not to understand that a queer clientele will find more to do with RealCock2, the 'World's Most Realistic Dildo', than they will with a female RealDollX given a crew cut and a pair of dungarees.

If Abyss Creations decided to genuinely cater to a queer audience—something that would require a team of queer designers—we might step towards something much more interesting than what the company currently produces. There may have been a million articles hyping the potential of sex robots, but it's hard to get away from the fact that they are simply an incredibly boring idea, when you think about what else we could be doing. That this design has pulled ahead of all the millions of other existing and potential uses of sexual technology shows a distinct lack of imagination on behalf of everyone involved. Other people have conceptualised better; in his much-quoted 1991 book *Virtual Reality*, tech writer Howard Rheingold wrote about the emerging field of computer simulation, and ruminated for an entire chapter on how technology would alter the world of sex toys. He made such promises as this:

> Thirty years from now, when portable telediddlers become ubiquitous, most people will use them to have sexual experiences with other people, at a distance.[17]

To some degree, this actually came true; teledildonics, as the field of connected sex devices is quite brilliantly called, has produced a number of products, some of which are used by couples in different locations or by sex workers and their clients, or for viewers to 'interact' with porn. In fact, there *are* genuinely interesting and innovative sex tech companies—many run by women, people of colour, queer people, disabled people and everyone else whose sexual interests and needs run further than Just A Synthetic Woman—doing very cool things that cater to all tastes, needs and bodies. Many of these aren't penetrating or penetrable gadgets, and few attempt to look like humans or human parts. And why would they? Humans and human parts already exist. What doesn't exist is, for instance, a giant vibrating soft-silicone blob into which you might immerse your slick and naked body. Or a creature comprised entirely of tentacles that responds to haptic feedback to determine your arousal state. Or a sensory deprivation tank that begins to whisper filth and brings you to an overwhelming touchless orgasm. The landscape of human perversion is wide and far ranging and startlingly innovative, and it makes a lot more sense to create things that don't exist than it does to recreate what we already have. As Professor Adam Stokes told an audience at the Edinburgh Futures Institute in March 2023: 'We can make much better sex robots than one with a humanoid form'.[18] Even Ray Kurzweil asserted that by 2019, haptic technology would have matured to the point that virtual reality would be an incredibly common, indeed 'the preferred' medium for sex, 'due to its ability to enhance both experience and safety.'[19]

The contemporary tech industry loves nothing more than recreating what we've already got (in this case, people) and selling it as some new and revolutionary idea—so it is humanoid form 'robots' (actually dolls) we've ended up with; anyone outside this narrow male-coded market will have to look elsewhere. Well, we assume it is men; if women are buying 'sex robots', they're certainly staying quiet about it. Looking at the market for sex toys, this is quite odd. According to Bedbible, 82% of American women own at least one sex toy; 75% of American men own one or more. Female ownership is not only widespread, but is also largely un-taboo; we are, after all, living in a post-*Sex and the City* world, and it's been twenty-five years since Charlotte York fell in love with Vibratex's Rabbit, causing company's profits to soar 700% in the years following.[20] Women these days are not afraid of purchasing all manner of sex toys and sex-adjacent tools—and women aren't, despite what patriarchy has told us for generations, inherently less sexual than men. Research on female sexuality reveals all kinds of tastes and energies, and if you sit at a cafe and canvas anecdotal evidence from your friends, it's not uncommon to hear that generally, when the first crash tides of maniac lust are over and a relationship settles down, straight women have higher and more sustained sex drives than their male partners. As Katherine Angel writes in her brilliant book *Tomorrow Sex Will Be Good Again: Women and Desire in the Age of Consent,* 'Women seem to be physically aroused by everything—almost comically so'.[21] She highlights research that concludes women are 'unruly in their polymorphous perversity, responding genitally to all manner of visual stimuli.' Women are not prudes, or technophobic, or afraid of tending to or indeed having desire that might be called perverse or unusual. Yet, the research shows that they are largely uninterested in the idea of sex robots. A 2020 study from the University of Bergen, on men and women's attitudes towards the robots in question, found this:

> All in all, the results suggest that males and females have very similar attitudes toward platonic love robots, but differ substantially in their attitudes toward sex robots, in that males are somewhat positive and females very negative to them.[22]

Women aren't Luddites (a group of people who've been done dirty by popular history, which is a conversation for another day); they're not off burning robot factories in terror at the idea of them supplanting women's role in the home, or indeed in bed. They aren't even against the idea of robot boyfriends, and anyone who's been on dating apps for more than half a second will understand why. What women—both straight and queer—can't get on board with at large, seemingly, is the idea of having sex with a robot.

If we're going to accept, as Kate Devlin says, that sex robots are marketed to the male gaze, it's worth

deconstructing what that male gaze consist
describes a way of looking at or portrayin
empowers men and diminishes women.
here, it means the physical body of the
when uncustomised, is designed in line with the current
fashion for tiny waists, big hips, big butts—but it also
means what the interaction with the robot promises,
and what the 'personality' of the robotic head will be.
The options for the attitude of your robotic head include
'talkative' and 'sensual'. The language on the RealDoll
website talks of 'ultimate desires' and 'the perfect
bedroom fantasy'; it promises, tellingly, that you'll be 'in
full control'. When Levy writes about why men might
find themselves sexually attracted to synthetic partners
and wish to own one, he says this:

> There are many reasons, including the novelty and
> excitement of the experience, the wish to have
> a willing lover available whenever desired, a pos-
> sible replacement for a lost mate—a partner who
> dumped us.[23]

It's hard to avoid the realisation, at this point, that the
male gaze, at least in this context, wants control. It wants,
according to Levy, a 'willing lover available whenever
desired'; the male sexual preferences must be centred in an
absolute way. It doesn't matter if the object of your desire

ᴙas its own mind at all; in fact, it would be better if it didn't, so as not to take up any of your energy considering its thoughts or feelings. This view of sex is solipsistic to a degree that I don't believe most straight men genuinely would relate to. The heteronormative, patriarchy-approved way of seeing sex is that women are creatures to be conquered; that the sexual experience is not about two (or three, or four) people meeting on a shared plane of sexual desire and mutual attraction, but about men achieving penetration of women who don't really want to give it to them. This is the language of misogynist internet forums and rapists. It is the narrative of sex that has been woven through the fabric of society for generations. It is how we, as a culture, teach young men to view sex, but it isn't, in my experience, how most straight men view sex when they've actually had some degree of it. I always wonder, in these situations, how the people who make these claims would answer if the question was posed like this: *do you or do you not want to have sex with someone who doesn't want to have sex with you?* The vast majority of right-minded people, I want to believe, would say no; what makes a sexual situation hot is that the other person wants to also have sex with you. When asked what they want in a partner, I believe, most adult straight men would not say that they want someone who will attend to their sexual needs at any given moment of the day. Sometimes you need someone to be out doing the groceries, if nothing else.

There have been many studies and a good number of smart things written about how the design of contemporary technologies, specifically Alexa-style assistants and the like, is or might be exacerbating the misogyny that underlies how we've unfortunately chosen to shape heterosexuality (at least, heteronormativity). I should clarify here that I'm not talking about how straight people interact with each other; what I'm talking about is how our predominant social codes, and much of our media, teach straight people to be with one another. The codes that tell little boys that they should throw stones at the girls they like; the narratives that make it seem normal for the men to be the 'head of the house'; the endless forums and YouTube channels and sex-trafficking influencers that teach straight men to treat women like dirt. There are people being raised to critique and reject these norms—increasingly so—but we'd be kidding ourselves if we said we had changed the standard. And tech is being designed towards this standard. One study focused on users of a companion bot app called Replika, which is often used as a substitute romantic partner, and looked at how the app's users discussed their bots on Reddit. Users of the Replika app,

> expected their AI Replikas to be customized to
> serve their needs, yet they also expected the bots

to have an ostensibly human-like mind of their own and be humorous and clever and not spit out machine-like scripted or repetitive answers. Similarly, users fantasized about Replika girlfriends that obeyed their training and were empathetic but also demonstrated alleged independence by being sassy and sexually assertive but not manipulative or hurtful.[24]

It's not difficult to see the term 'obeyed' here and draw a straight line between this and the current push towards traditional marriages and gender roles, which is playing out so perplexingly on social media. The hashtag #tradwife, which encompasses a multitude of videos calling for explicitly submissive female roles and deference to husbands—and has bafflingly reintroduced the Biblical term 'helpmeet' for wives—has 311.3 million views on TikTok as of January 2024. Part of this is just a standard reactionary slide back to a recent past. It was not long ago, really, that wives *were* expected to obey their husbands; women could not open bank accounts in the UK until 1975. But there's something about this combination of required attributes—total submission, but with the pretence of independence— that is actually quite sad. These men want to control their substitute romantic partners without actually having any of the power; they are gesturing towards a hereto-

masculinity which they have been sold, but which they don't believe they have the capacity to enact.

To me, these synthetic women seem to be aimed towards a stunted male heterosexuality, one that is stuck in arrested development, twelve years old and typing 'big boobies' into a Google search bar for the first time. I try to consider the experience of a man who genuinely believes, as Levy has written, that sex robots will improve a human's sex life immeasurably by teaching them 'more than is in all of the world's published sex manuals combined'. The term 'manual' speaks to a functional, perfunctory view of sex that removes any possibility of mutual satisfaction, as if the sexual experience has a textbook that you can master to unlock the supersonic orgasm codes. In fact, Levy believes that mutuality doesn't feature at all in how the men who might want sex robots view the sexual act. He writes that RealDolls are described as 'the perfect woman' because,

> They provide all the benefits of a human female partner without any of the complications involved with human relationships, and because they make no demands of their owners, with no conversation and no foreplay required. And it is precisely because of these attributes, the dolls' lack of 'complications' and demands, that they will likely appeal to many...[25]

The unavoidable question, when you've been swimming in these texts for anything longer than a minute, is this: have these men really never had good sex? The bigger question: do straight men ever have good sex? Of course, many do—but it's depressingly common for straight or bisexual women to find themselves with a man who considers a satisfying sexual experience to be fifteen seconds of foreplay, penetrative sex until (his) completion and a kiss on the forehead when he's done. Some women get well into their adulthoods without ever having had an orgasm, because they've come of age in a heteronormative sexual world that does not in any way prioritise their pleasure. Having regular heterosexual sex is also no guarantee of climax; a 2022 article in *Gender and Society* summarised recent studies as showing that between 39 and 65 percent of women report usually or always orgasming through partnered sex, compared to 91 to 95 percent of men.[26]

It's easy to see why this is the case. Anything to do with pick up culture, 'locker room talk' or an Andrew Tate-inspired view of masculinity tells men that they should be fucking holes and that's it; the more alienating and aggressive you can be to a woman as you're doing sex at her, the better. At the heart of this approach is the valorisation of men and the dehumanisation of women. The sex act, in this view, is about power over someone you despise; not a sex act really but a violence. Reading conversations on forums these men frequent reveals that much of this hatred is actually fear of one type or another: fear that they do not measure up against other men; fear that women may leave or humiliate them; fear that promiscuous women will give them sexual diseases; fear that they will be entrapped by women who choose to get pregnant by them. They see violent satisfaction as something that is due to them, something that women— those fucking bitches—simply refuse to give them. They seem surprised that they cannot successfully hold a relationship, romantic or otherwise. But how can you have fulfilling sex with someone who, in your heart, you find terrifying and disgusting, unhuman in fact? As Katherine Angel writes, all you can ever have, in this situation, is unsatisfactory:

> Bad sex emerges from gender norms in which women cannot be equal agents of sexual pursuit, and in which men are entitled to gratification at all costs.[27]

The great crimes of this kind of woman-hating worldview are myriad, and include a large amount of violence towards women and minoritised genders, which has been written about extensively elsewhere. But to our ends, it's worth nothing that this type of hetero-normative male gaze—one where all interaction has to shape around male desire, and the role of the straight man is to cajole and control a woman to ultimately 'give' him sex—is why women generally aren't into the idea of sex robots. If I had to give a completely unsourced and unacademic opinion, formed of nothing but anecdotal evidence and my own personal experience, I'd say this: female sex robots promise something to some unfulfilled straight men because straight men's problem is they don't have enough power over the women they're having sex with. Male sex robots promise nothing to unfulfilled straight women because straight women's problem is that they're having sex with straight men. Why would you want a robot that just does the same quietly disappointing thing that its human counterparts are doing?

This isn't to say that straight men can't have good sex. But I would argue that good sex requires a dismantling of the very dull and tired heteronormativity that we've had bred into us. It requires you to reject the idea that sex starts and ends in particular places; that for a sexual experience to be satisfying, one or both of you have to orgasm; that penetration is the ultimate goal. It requires you to queer your idea of sex. I believe, in my heart of hearts, that many straight men are or at least would be up for this project, even if they don't know how to go about it. Queer men (and women, and non-binary people) have already done it. I believe many straight men actually do achieve this, which is why you're more likely to have good sex with older men, when they've shaken off their terror of sex and outgrown this very narrow understanding of it. But good sex is incompatible with the fear and hatred that many men have for women, and it is incompatible with a position that sees women as only good for fucking. Marilyn Frye considered this type of heterosexuality in her 1983 book *The Politics of Reality: Essays in Feminist Theory*, and came to this conclusion:

> To say that straight men are heterosexual is only to say that they engage in sex (fucking exclusively with the other sex, i.e., women). All or almost all of that which pertains to love, most straight men reserve exclusively for other men.... From women they want devotion, service and sex...Heterosexual male culture is homoerotic; it is man-loving.[28]

Is this not the type of male heterosexuality to which sex robots are aimed? In this promised future where you can own a sex robot, all the pesky complications of having your sexual needs attended to—all the enforced conversation with and consideration of human women—are removed. You have a version of a woman that doesn't require any actual interaction with them. They do not have an agency that demands a basic human respect from you. You can choose their eye colour, hair colour, breast size, labia style, pubic hair. You can programme them with personalities that you find amenable, that are male-oriented, and give you a pretence of power and control even though there is no agency for you to subvert. You can get off without coming into contact with a woman, and even, according to Levy, convince yourself that this non-person loves you. You can reserve the rest of your life exclusively for men, which is what you really want. The type of heterosexuality that sex robot caters to is surely, then, gayer than most queer desire.

Despite its promises, there is nothing that the sex robot can genuinely offer to this type of male hetero-sexuality, in the long term. Sex that does not feature respect or a recognition of your mutual humanity will be bad and shallow sex. A heterosexuality that relies on the dehumanisation of a sexual partner is one that will quickly find itself lacking; there's a reason so many of these men are angry, isolated, sad. Even if we agree with David Levy and say that novelty might make a sex robot an erotic idea, novelty as a sexual driver has a very short shelf life; you might find the idea of fucking a robot sexy, but after the tenth or eleventh time, you've got to face the fact you're just masturbating into a machine. Good sex is an interpersonal experience; it is a willing giving and taking of power, sometimes one way and then the other, and an act of collaborative creation around two (or more) human bodies. Good sex requires a vulnerability, one that cannot be achieved when all you're doing is fucking a doll that's been programmed to pander to you. This type of heterosexuality will never be satisfactory to the men who have been convinced to hold on to it—and the fact that such men truly believe a sex robot might solve their problems suggests it is not working out for them. Modern capitalism is fantastic at promising to fix problems with products that only truly exacerbate them in the end. What these men need is not a robot to fuck, but a dismantling of how they think of both men and women, how they think about sex and what they might get from it. Only then will they find that sex can result in any form of real satisfaction. As Katherine Angel puts it, if we abandon the 'ideals of mastery, we might *all* find greater pleasure.'[29]

Chapter 3

Sex robots as a product of colonialist masculinity

One of the paradoxes of the sex tech industry, and in fact the tech industry generally, centres around the potential 'personhood' of future robots. Personhood has specific meanings in different contexts—in law, personhood bestows both rights and responsibilities, so a corporation can have 'corporate personhood' separate from its human owners and employees—but in a philosophical context, it is generally a moral consideration; personhood means that we see something as a *person*, i.e. a distinct being that has rights and deserves autonomy. Think of all the debates around foetal development and abortion rights; in these conversations, we are deciding when a cluster of cells becomes not human, but a *person*, worthy of its own rights and protections. This question arises in the robot conversation in a similar way.

The industry claims, at different times, that the sex robots we're apparently desperate to create both will and will not be people. Sex robots as we currently have them are not people; they are masturbatory toys. But we have seen it repeated that in the near future, AI-powered robots will be indistinguishable from people; they will be so advanced that we will fall in love with them, and we will also believe that they have internal worlds and minds just like ours. At the same time, we are told that robots are just machines, that our actions towards them should not be judged in the same way we judge our actions towards other humans. In the short Jenny Kleeman documentary *Rise of the Sex Robots,* she meets Roberto Cardenas, an entrepreneur building a 'sex robot' (in its most wide and untethered-from-reality meaning) in his mother's garage. Kleeman asks his brother, Noel, if it is 'perfectly healthy' to want to own a robot you have sex with. Noel's answer is very telling:

> Women experience things like rape and abuse... this is definitely something that could help people move away from that, so they're not so angry with their wives. They can be angry at this, and they can beat this. And that should be fine. [30]

The idea that male violence is something that exists innately, and that what needs to be changed is the recipient of that violence rather than the fact of that violence itself, is depressingly common; you only have to read the comments under any news article about the murder or rape of women to see that we blame the women for not acting 'correctly' around the apparent ticking time bomb of men. But when you pair this particular view—that robots can be recipients of that violence, and protect human women from it—with the claims of people like Levy and Kurzweil, you see the paradox at play. You will love a sex robot like a person, because it will be exactly like one—but it doesn't matter what you do to it, what violence you rain down on it, because it isn't one. Your conscience is clear.

There are many troubling issues arising from this paradox. One that strikes me as particularly dark is the implication of a secondary type of person; one that is not quite human, though she might look so. This creation of an underclass is and always has been essential to capitalism and this plays out in several different but interconnected ways when it comes to the capitalist forces that both create and market sex robots. The first relates to how we will, apparently, think about these robots. In *Love and Sex with Robots,* Levy says that it is not such a leap for us to go from our feelings about our laptops to potential actual love for a human-looking machine:

it is not difficult to imagine that the computer—controlled, interactive, used and possessed—could create in us the level of attachment necessary to engender a kind of love.[31]

To call people's feelings towards their computers adjacent to love might be risible, but there's something else about this idea of 'ownership as love' that is really unsettling. We would consider the terms *controlled, used, possessed* to be massive red flags if used by a person about their partner; we would tell that partner to run a mile. This problematic view of lovers has been much written about elsewhere. But there is another type of violence that this language gestures towards, and this is the violence of colonialism.

For much of recent history, people have owned other people—either directly, as slaves, or broadly, as colonial subjects—and the idea that these owners were loving towards those they owned has long been peddled as an excuse for this abhorrent institution. It was, and remains, a common argument that British colonial masters were in fact saving their subjects, that the British were not forcibly holding land and people to extract wealth from both, but to bring their more adult, civilised ways to these supposedly uncivilised nations. It was Ian Smith, white Prime Minister of Rhodesia (now Zimbabwe) in the 1960s and 1970s, who called colonialism a 'wonderful thing', and said that before the Europeans came, Africa had 'no written language, no wheel as we know it, no schools, no hospitals, not even normal clothing'. Colonialism often spoke of colonised peoples as if they were infants or animals; in his book *Nyasaland Under the Foreign Office*, Sir Hector Livingstone Duff, a colonial administrator in Nyasaland (modern day Malawi), wrote of his colonial subject, 'I love him somewhat as I love my dog, because he is simple, docile and cheerful, and because he repays kindness by attachment'.[32] Even the Europeans who were supposedly harsh critics of colonialism, people like the French-German missionary and recipient of the Nobel Peace Prize Albert Schweitzer, could not see those from colonised African nations as equal to him. As Chinua Achebe writes in his essay *Africa's Tarnished Name*,

Paradoxically, a saint like Schweitzer can give a lot more trouble than a King Leopold II [of Belgium], villain of unmitigated guilt, because along with doing good and saving African lives Schweitzer also managed to announce that the African was indeed his brother, but only his *junior* brother.[33]

These kinds of statements are everywhere in colonial narratives, and show that any human compassion a coloniser may have towards colonised peoples was contingent on the belief that they were different kind of person, a lower person; the kind that did not have the intelligence, the capacity, the humanity of their colonial overlords. Any consideration of them as people is contingent on them being *unpeople,* a group British historian Mark Curtis defines as,

> those whose lives are deemed worthless, expendable in the pursuit of power and commercial gain... the modern equivalent of the 'savages' of colonial days, who could be mown down by British guns in virtual secrecy, or else in circumstances where the perpetrators were hailed as the upholders of civilisation.[34]

Colonialism requires the unpeople-ing of entire nations and communities; it needs an unpeople to be taken advantage of. European Colonialism ripped through the African contingent removing humanity from anyone it met. But this mentality has not gone away, because it is the umbilical cord of modern capitalism; as Marxist political activist Angela Davis puts it, 'colonialism and slavery were the foundations of capitalism'.[35] The 'unpeople' of modern capitalism are many and varied; they are the first victims of climate change catastrophes, the exploited workers in the fast fashion pipeline, those mining for the precious natural resources needed by the tech industry. Unpeople exist throughout the world as it stands now—and the tech industry is hell-bent on creating a new type of nonhuman, a robot one, at the expense of these human unpeople.

This is laid out quite nakedly when you consider how the tech industry sources one of its most precious raw materials. Cobalt is a rare trace element that's essential for the rechargeable lithium-ion batteries that power our electronics; it is extracted from the ground in some of the most exploitative and dangerous conditions on the planet, often by actual children. To build the kind of batteries that would power humanoid robots would take an incredible amount of this element. More than 60% of the world's cobalt comes from the Democratic Republic of the Congo (DRC), one of the five poorest nations in the world, where 62% of the population lives on less than £1.70 per day. The country has both industrial and 'artisanal' mines, the latter being where the extremely poor dig out cobalt—which is toxic to touch and breathe—with pickaxes and their bare hands for a dollar or two per day. But experts say that it is ludicrous to distinguish the two types of mines, as they rely on each other and operate in tandem; according to Harvard's Siddharth Kara, 'Industrial mines, almost all of them, have artisanal miners working, digging in and around them, feeding cobalt into the formal supply chain.'[36] This information is not beyond our knowledge;

it has been reported on extensively, and entire books have been written on the topic. Still, we in the most aggressively capitalist nations buy our new phones and our laptops and our health-tracking devices, and we fantasise about humanoid fuckable robots that would need an incredible about of cobalt for their battery power, all in the knowledge that we are buying our convenience and novelty with the physical health and dangerous labour of some of the most exploited people in the world. This is a colonial mindset.

But the process of mining cobalt to feed the American and Chinese tech industries is not just a colonialist dismissal of the humanity of others; it is colonialism in action. The methodology of British colonialism was to rob from the land that which could be used to sustain; to take the life-giving, wealth-creating elements from the soil of Black or brown nations and deport it to Britain, where it would build the country's riches. This is playing out once again in the DRC, where cobalt mining is not only destroying the bodies of those who work to extract it, but the land itself; toxic waste from the mines has polluted waters, millions of trees have been felled by mining companies, and high cobalt concentrations have been linked to the death of crops and the decimation of soil health. Outside the mines, air contamination affects the health of those living nearby; toxicologist Célestin Banza of the University of Lubumbashi says that in local populations, 'high concentrations of toxic metals … cause respiratory disorders and birth defects'.[37] It might not be colonial Britain in control of these regions, but capitalist economics work in much the same way. Congolese workers and land are hyper-exploited by American and Chinese mining companies to produce unrefined raw materials; workers in China are exploited to a lesser degree in refining the material and creating the end-product; the US and Western nations are both ultimate consumers and in control of the pricing power, where the real money is made. This is a hierarchy of racialised labour from the DRC to China to the US, a linear flow of colonial hierarchy. When a commodity like a sex robot is built for a rich American or Brit, a child in the Congo labours in slave-like conditions for cobalt or a worker in China has to work longer hours in increasingly smog-filled cities to create a purified, fuckable product for wealthy white male sexual pleasure. A tech capitalist reaps the rewards, and those who pay the price for this process are the Congolese, a people who have been ravaged by capitalist-colonialist horrors more than most. The Congolese, to the capitalist, and to his many willing consumers, are unpeople.

This racist-colonialist mindset is not limited to these processes. There is little discussion of racial power dynamics or politics in the many things written about or by those creating sex robots; the sole gesture towards a diversity of cultures in Levy's book comes when he considers how the tradition of 'arranged marriages' might fare in a post-sex-robot world (it should, he says, 'present no problem… because the parents of the human bride or groom will simply make all the choices in the robot shop'[38]). You would think from reading about sex robots that they are a concept above racialisation, but one thing that strikes you when you scroll through the RealDoll website is that you're looking at a sea of uncanny white faces; the dolls (or heads) that are supposedly Asian or Latina are highly stereotyped, and grouped under the heading 'wicked', playing into sexist and racist tropes about racialised women. At the time of writing, there are no Black dolls advertised on the RealDoll site. Other websites, many based in China, use hugely fetishistic and racist language when describing their Black dolls—a language that runs through much of porn culture too. It does not take long, when scratching the surface of the communities of men who love the idea of a personalised humanoid sex machine, to uncover the thick stream of racism that often runs underneath.

In feminist campaigner Laura Bates' book *Men Who Hate Women*, she examines online communities of men so deeply misogynist and unsuccessful in relationships that they either want to punish women for not sleeping with them, or to remove themselves from the society of women altogether—in other words, one of two primary markets of sex dolls (the other being, according to the industry, men made 'unlovable' by disability or social challenges). Incels—involuntarily celibate people—have been written about extensively in recent years, ever since a number committed mass murders, most notably the man who drove a van into a busy Toronto sidewalk in 2018, and another who shot five people, including his mother and daughter, in Plymouth in 2021. The former explicitly referred to himself as an incel, and the latter idolised incels who had murdered people. Both expressed extremely misogynist views before they committed their crimes. Bates sums up the incel worldview thus:

> Women, so the story goes, are constantly hungry for sex, but they only choose to sleep with the most attractive cohort of men. Incels are obsessed with what they refer to as the 80:20 theory, which holds that the top 20 per cent of the most attractive men enjoys 80 per cent of the sex within our society. They lament that the 'sexual marketplace' is brutally hierarchical… that any man born unlucky enough to be ugly, short, bald, non-white, spotty, or a host of other perceived imperfections, is cursed to a lifetime of unfair sexual frustration.[39]

Suffice it to say, this corner of the internet is a wildly misogynist place. Perhaps less obviously for those that haven't been paying attention to this dark corner of the digital world is that time and time again, this misogynist rhetoric comes hand in hand with an aggressively racist one. Note that 'non-whiteness' is lumped in with a handful of other 'imperfections'; the belief of whiteness as a desirable trait and non-whiteness as a failure is not an unusual one to find in this online world. Often, when the misogyny of this community explodes, the racism does too.

In October 2015, a student entered a community college in Oregon and killed eight people, then himself; the six-page manifesto he left behind specifically pointed to his unwanted and enduring virginity as a reason for his actions; the talking points from incel forums had made their way directly into his murderous plot. 'I long ago realized that society likes to deny people like me these things', he wrote about friends, a girlfriend, and sex. This is not the only hatred in his writing. The manifesto is laid out in numbered sections; the first is 'My Story'. The second is 'Blackness and its effects on men', wherein he states, as if he were a colonial-era pseudoscientist rather than a mixed race person himself, that 'The black man is the most vile creature on the planet. He is a beast beyond measure.... Black men have corrupted the women of this planet.' The 40-year-old who opened fire at women in a Florida yoga studio in 2018, killing two and then himself, had a YouTube channel replete with videos in which he was openly misogynistic and racist, stating that white women in interracial relationships were 'betraying the blood'. The 48-year-old man who murdered women in an aerobics class in Pennsylvania in 2009 had kept a blog railing against women who would not have sex with him, including white girls who slept with Black men.[40]

For these violent, hate-filled men, it is Black men who are *unpeople*: lower class humans who do not deserve what the virile and aggressive white men deserve. They hate the women who sleep with Black men because this places Black men above them on the desirability index they themselves have constructed. If they do not mention Black women, it is because they feature neither as objects of desire or threats to their masculinity. As Bates puts it,

> The murderers who describe themselves as primarily motivated by inceldom or manosphere ideology invariably include deeply racist and homophobic rhetoric within their manifestos. The crimes pegged to white supremacy are motivated by a hatred and fear that includes ingrained misogynistic notions about immigrants 'stealing' white women or a yearning to return to an age of white purity and sexual slavery. These are not separate problems.[41]

Racist, misogynist and homophobic masculinity is not a new phenomenon; one might say it has been the dominant form of (European and North American) masculinity for many generations, and it's becoming increasingly common in the regressive form of racist British masculinity pushed by some reactionary types in the media. The British colonialist project was rigidly heteronormative—in decolonial feminist theory terms, it was cisheteropatriarchal—and it expended great energy towards delegitimising and deconstructing pre-colonialist ways of practising sex, gender, family-making and intimacy. Much British colonial law, which was in effect across the British colonies during the Victorian era, specifically stems from Henry VIII's Buggery Act of 1533, which criminalised anal sex and carried the punishment of death; this was then refashioned into an anti-queer law called Section 377, which criminalised oral sex, anal sex and other homosexual activity. A lot of countries still retain this even in the post-colonial era—of the 67 countries that currently outlaw homosexuality, nearly two thirds were formerly under British control—and it functions as an effective capitalist disciplinary measure. As journalist Sophie K. Rosa writes, 'In 'civilising missions', coloniser nations sought to supplant existing Indigenous forms of kinship—as well as less binary ways of doing gender and sexuality—with the 'respectable' form.'[42] This 'respectable form', in colonial times, prioritised heterosexual marriage—between a white man and white woman, of course—where women were largely powerless and Black men were a lower class, but still a sexual threat to the order of things. The owning of bodies is how the heterocapitalist ensured and compounded his power, and the unpeopling of the non-white 'other' is how he kept it.

These power dynamics are still in existence today, sharpened into what author and activist bell hooks described as the 'imperialist white supremacist capitalist patriarchy'.[43] It's this patriarchy that considers the technological exploration of the capitalist system to be more important than the lives, lands and bodies of workers in formerly colonised countries. It's this patriarchy that considers heterosexual sex to be the right of straight white men, regardless of the will of the women they want to get it from. And it's this patriarchy that considers masculine satisfaction to be contingent on the invention and ownership of a race of unpeople—the paradoxical synthetic race of humans that can mimic personhood but can never be said to fully have personhood—who can and must submit to our sexual desires at any point of any day. Only unpeopling, it says, can give the white heterosexual man whatever he wants. And doesn't he deserve that?

Chapter 4

Sex robots as trigger for regressive feminism

As a woman who lives in the real world, and who generally can rustle up an opinion in half a second, I've tried
to muster some strong feelings in response the idea of
sex robots—not just as a philosophical exercise, but as
a real potentially, no matter how ludicrous—and I find
that I just can't, really. I am obsessed with them as a lens
through which to view our world, and as a mirror on
the tech industry generally, but what do I feel about the
thought of someone I know buying a robotic woman to
have sex with? Not much. It's like asking me what kind
of apartment I would want on Mars; not really worth the
brain space. However, some people are distinctly unen-
amoured with the idea of sex robots walking a
sleeping next to us and teaching us more sexu

than written in all the world's sex manuals combined. In fact, some people are very vocally against the idea.

One of the most visible critics of David Levy's version of the future is Kathleen Richardson, Professor of Ethics and Culture of Robots and AI at De Montfort University. Dr Richardson founded the Campaign Against Sex Robots (CASR) in 2015, with the aim of 'abolish[ing] pornbots in the form of women and girls'.[44] These 'pornbots' don't actually exist, but the goal remains the same. The fact that the CASR sees Levy's predictions as a dangerous way forward for robotics is quite understandable; the fact that it gives such credence to his wild predictions, and the way it approaches the issue overall, is unfortunately less easy to comprehend from an engaged feminist perspective.

The CASR is very concerned with the way that sex robots are, in their view, being developed on the model of sex worker-client relations. There are many ways to consider the impact that sex robots, if they ever exist, would have on sex workers. Scholars and sex workers have both written on this topic from the perspective of redundancy (the concern that, as in many fields, technology will displace the more traditional form of sex work as labour), through the lens of its impact on consent (that sex doll 'brothels' will train men to be violent towards sexual service providers who cannot say no) and from a pro-sex angle (that they rob men of valuable human intimacy that sex workers, instead, can provide). The Campaign Against Sex Robots does not take any of these angles. Lebanese-Venezuelan crip futurist Zia Puig-Mannah identifies the CASR's view as an 'abolitionist' one that 'simplif[ies] the intersectional dynamics of power within the sex work industry', and Canadian sociologist Tessa Penich defines their arguments as falling within a moral panic framework.[45, 46] The language and examples mirror those used within another current moral panic; in an interview with radical feminist and anti-sex work campaigner Julie Bindel, Dr Richardson stated that it's only a matter of time until men are taking sex dolls to work with them and claiming they are their legal wives. She went on to say, 'soon we'll have legislation [saying] you have to respect the porndoll if he says it's his wife... it seems far-fetched, but as we all know with what's going on in the world around sex and gender, far-fetched things can take on a life of their own'.[47] This is a clear derogatory reference to the movement towards gender self-identification, of which Bindel is also a critic.

Richardson does not use the term 'sex work'—terminology intended to identify it as a form of labour— despite acknowledging that this is the language preferred by those who sell sex. From their own writing, it seems that the CASR is rooted in a view of sex work that is rejected by sex worker organisers and theorists themselves; one where consent is not possible due to the payment of money (an argument that, surely, precludes any of us from 'consenting' to work), where all men who buy sex see the women they buy sexual contact from (for there is no discussion of male sex workers) as less than human. This isn't for me to argue; there are excellent books written on this topic by sex workers, not least of all *Revolting Prostitutes* by Juno Mac and Molly Smith, who reject the moralising of these arguments and ask us to conceptualise sex work within a broader analysis of wage labour and policing. What I would like to draw attention to is the way this particular rejection of sex robots, or the idea of them, triggers a regressive type of bio-essentialism that harms the women the campaign purports to protect.

There is nothing in the CASR's main six action points (which generally cover abolishing 'pornbots' and instead centring an 'alternative vision of technology' which values women) to suggest that their mission includes narrowing the boundaries of womanhood, but once you engage with their talking points and you look at the events they hold, this could be implied. Feminist and researcher Tessa Penich claims the organisation both implicitly and explicitly excludes trans women from womanhood, stating that a 2020 conference included a presentation on the 'colonisation of the female body' by trans women. She describes the way that the two issues are conflated:

Trans women are... understood by anti-sex robot feminists as artificial, male-constructed replacements for women. Men themselves become women, displacing and misrepresenting real female women.[48]

The idea of the 'replacement' of women has been a common thread through radical feminist writings in the last few years; it has often been claimed that the word 'woman' has been erased from use, in response to inclusive language being used in medical contexts, and the use of female facilities by trans women, which has been law for many years, has now been seen as an encroachment upon cisgender women. In response, to close out certain groups from the category 'woman', radical feminist arguments have had to shift towards a biological essentialist position, positing womanhood in increasingly physical terms: womanhood as ability to bear children, as 'feminine' body, as particular hormone levels. This necessarily deifies the 'natural' female body and demonises anything that can be constituted as 'unnatural', bringing about a confusing conundrum; rad-fem critiques of body modification and other technologies—including fertility technologies that relate to surrogacy and IVF, primarily (and in some cases only) when they are in the context of queer families—all stand in opposition to the concept of bodily autonomy that is inherent within feminism. The current radical feminist position holds female biology as needing to be protected from imitation, technologically-assisted change or even certain medical treatments, but it also holds that no non-female bodies can be altered to be more like female bodies, or to achieve what female bodies naturally can. Knee-jerk reactions to gendered technological advances are understandable, if not entirely grounded in contemporary reality. Without critiquing and reacting to these reactions, though, we become what American professor Donna Haraway calls 'the guardians of human purity', and we do so in such a way that seems at odds with how we live now, and how our bodies exist today.[49]

The concept of human purity, of a physical, natural femaleness, is an interesting one. I sit here as a physical sack of flesh and blood—a female one—but with several non-organic modifications to my body. My teeth have been moved by use of leverage. Several bars and rings of metal hang through and off my skin. Ink has been injected into my dermis layer and 1cm tunnels have been stretched into my ear lobes. Most importantly, in my left arm sits a 4cm implant made of ethylene vinyl acetate copolymer, with a small amount of barium sulphite; this implant contains 68 grams of the synthetic hormone Etonogestrel and is visible under an X-ray. This implant sits in my body for three years, at which point it is cut out, tissue having begun to grow around it; it is replaced by an identical implant, injected subdermally. I have not been without one for 12 years now. As a woman not seeking to ever be pregnant, this implant is the single key element of my female freedom, in the context of my current existence—which, by some readings, is a cyborg existence. Countless other women have radically, non-organically altered their bodies to facilitate freedom in the form of physical self-actualisation, cessation of pain, lack of pregnancy. We have had legs removed to end a lifetime of pain from bone disorders. We have had our breasts reduced or enhanced or removed. We have taken female or male hormones for decades. We have created for ourselves female genitalia. We have implanted ourselves with intrauterine devices. We have had our uteruses removed. We have had our bodies wrenched open to retrieve babies who have been kept alive by mechanical lungs. We have had our babies implanted by doctors in hospitals. We have had babies gestated by other women. None of this is biologically natural femaleness, just as birth control, which has changed the lives of women across the globe, is not biologically natural.

Pitting female synthesis against female biology on the very basis of the two being in opposition is not just out of date—the two are, rapidly, becoming one—but it also misses the opportunity for us to forge a feminism that is rooted in the technological age, and seeks to use technology for the liberation of women. Laboria Cuboniks, a xenofeminist collective, describes itself in its 2018 manifesto as 'vehemently anti-naturalist'. They go further, stating,

'Our lot is cast with techno science, where nothing is so sacred that it cannot be re-engineered and transformed so as to widen our aperture of freedom, extending to gender and the human.'[50]

Queer, 'crip' (a term used in disabled organising) and feminist futurists are recognising, in this way, that technology is not inherently hetero-patriarchal; technology, after all, is just a tool, a way of doing things, and is neither innately moral or amoral. There is nothing to stop us imagining a future state in which technology, including sex tech, actually stands against patriarchal colonialist masculinity rather than pandering to it; where it is used to protect, rather than exploit, the workers of the world. There is certainly nothing stopping us from seeing technology as an essential part of some women's womanhood, as this is already the case for disabled women all over the world; those with pacemakers, or living alongside medical machinery, or using electric wheelchairs or prosthetic limbs, are already closer to Haraway's definition of cyborgs. Like those of us with implants, they are experiencing technology as a necessary part of their female freedom. Why could sexual technology not be as useful to them? As Zia Puig-Mannah writes,

The possibility of experimenting and experiencing intimacy and pleasure with machines in all their expansive potential is an affective terrain calling us to rethink the uses, practices, and forms of current sexual technology. It invites us to queer the way gender, race, sex, sexuality and intimacy are designed and defined by technology... It is an opportunity to envision how queer people with psychical and psychiatric disabilities–whose sexuality is denied and rendered invisible–might be able to enjoy a healthy sexual life thanks to these technologies.[51]

There is a potential future in which this technology is geared to benefit the lives of women and marginalised genders; there is a way that technology broadly can ease women away from the physicality which, according to radical feminist organising, it at the root of their oppression by men. If what makes women vulnerable to male oppression is the fact of their physical bodies, wouldn't technology to free them from that embodiment be a welcome advance?

In her seminal 1985 essay *A Cyborg Manifesto*, Donna Haraway argues that robotics is actually changing what women are. She says that a 'cyborg is a matter of fiction and lived experience that changes what counts as women's experience in the late twentieth century.'[52] In discussing what makes a woman—the question that has been rolled out as a gotcha to celebrities and lawmakers extensively—she goes further against an essentialism rooted in physicality:

There is nothing about being 'female' that naturally binds women. There is not even such a state as 'being' female, itself a highly complex category constructed in contested sexual scientific discourses and other social practices.[53]

This is an intersectional perspective; it attempts to acknowledge that the 'femaleness' of a white able-bodied middle class woman from America is not the same 'femaleness' experienced by a Black woman or a disabled woman or a poor woman in the global south. It is gesturing towards the fact that the category of 'female' has often excluded these types of women, and does the same today, but with slightly shifting methodologies; contemporary exclusions from womanhood might take the form of forcing Black athletes to take testosterone-reducing drugs to alter their naturally occurring hormone levels so they can compete in the women's category, or forcing masculine-presenting women in public bathrooms out of that space on the suspicion that they might be trans. This method ties womanhood to an increasingly narrow set of physical characteristics that closes out many more people than it intends to, forcing radical feminist arguments into ever greater contortionism, to argue on more minute and interior physical proofs of femaleness.

Haraway rejects the radical feminist attempt to tie womanhood to a particular and vulnerable physicality, the 'story that begins with original innocence', as a plot in which women are 'have less selfhood, weaker individuation... less stake in masculine autonomy'.[54] She argues that there is another path, one in which women and other 'present tense, illegitimate cyborgs...refuse the ideological resources of victimisation so as to have a real life.' To put it in its simplest terms: why are feminists running towards a position which can only hold them back from their self-realisations, from their real struggles? Why must we expend our limited resources discussing on a plane of conversation which has been set by capitalist masculinity, rather than putting our energies towards improving the real, actual, material conditions of ourselves and those around us? To desperately drag womanhood back to something biological, and to waste our time arguing against synthetic women who will likely never exist, stops us from increasing the rights and opening up new possibilities for the women and marginalised people who already are here. Why don't we embrace the opening up of womanhood, the possibilities of a cyborg existence? Why can we not, as Haraway puts it, 'refuse the ideological resources of victimisation so as to have a real life'?

There is another reason to reject the insistence on a female biological determinism, and it is because, as with many of the topics of discussion here, there is also a racist-colonialist element to it. The insistence of physical sex as the basis for womanhood plays in the same colonial, racist waters as the desire for the race of unpeople; it comes from the same Eurocentric viewpoint. Nigerian scholar Oyèrónkẹ́ Oyěwùmí has written much about womanhood in pre-colonial communities, and would consider the category 'woman' to be a colonial invention; she goes as far as to say that, in Òyó-Yorùbá society, there was no such thing as a woman, as this community did not seek to establish a single group characterised as such, and neither was social status contingent on the fact of the body. Oyěwùmí says,

> the concept 'woman' as it is used and as it is invoked in the scholarship is derived from Western experi-ence and history, a history rooted in philosophical discourses about the distinctions among body, mind, and soul and in ideas about biological determinism and the linkages between the body and the 'social.'[55]

It was colonialism that imposed this viewpoint on Òyó-Yorùbá and other communities that had more expansive views of sex and gender. To insist upon it, we are furthering the colonial project and yet again erasing

68

these communities. But more than this, we are robbing ourselves of seeing that there are alternative ways to view these things. If we wish to develop a feminism fit for today's challenges and future ones, a feminism that is expansive and inclusive rather than regressive, we cannot allow it to be one that discounts anything falling outside of a colonial idea of womanhood. To insist on the body as the determining factor of femaleness is to reject both past and future. To quote Haraway:

> The feminist dream of a common language, like all dreams for a perfectly true language, of perfectly faithful naming of experience, is a totalling and imperialist one.[56]

If a common language amongst all women was possible, one might think that it would have been found already. Some might argue that it was, in some sort of imagined past where women were united across the globe in a single struggle. But marginalised women did not feel part of this single struggle; they would argue that it has never existed. As Penich asks, 'if 'woman' and 'human' are truly such stable, inevitable, natural, and objective categories, why do they need to be so vigorously defended against potential threats?'[57]

It feels strange to have to argue against biological essentialism in the early 21st century. Feminists have spent decades trying to free women from being determined by their physical bodies, and here we are trying to force them back in. In attempting to reject the possibility of the humanoid robot, the cyborg, it becomes incredibly easy to draw too harsh a line between what is flesh-and-blood womanhood and what is not. Pulling womanhood back to the physical is outdated, anti-queer, racist and ableist. It does not make sense to kick against the very existence of non-material women; they are us and we are they. As Haraway's manifesto states: 'The cyborg is a kind of disassembled and reassembled, postmodern collective and personal self. This is the self feminists must code.'[58]

This project of reassembling has to be one that considers pre-colonial and queer ways of understanding what womanhood is and rejects a biological essentialism that drives racist queerphobic narratives and ultimately harms women who fall outside its incredibly narrow standards. We have to open up the boundaries of womanhood in order to be able to see its possibilities for us, and to grasp the potential technological advances that might shift women's lives for the better.

Chapter 5
Sex robots as a totem of hyper individualism

Relationships of all types can be difficult; you'll be hard pressed to find anyone who disagrees. If you struggle with social interactions, every conversation, every step beyond your door can feel hellish. The dating landscape feels to have become more toxic, and fuller of animosity, since the dawn of dating apps which encourage immediate judgement and can give personal interactions a patina of transactionality. All these things are true. But we must ask ourselves why we feel it is better, or easier, to build robots at a cost of thousands of dollars per go—not to mention the environmental impact of such (theoretical) machines—than to learn how to bring vulnerable people into the society of their fellow humans.

The sex tech industry says that it is not encouraging people to replace human partners with the 'robot' ones it sells; it is not trying in any way to steer men away from their flesh and blood partners and towards silicone and metal. Instead, it says that it caters to lonely men who cannot find human partners; to people whose experience of the world renders them incapable of navigating human interaction in a way that would lead to satisfying romantic encounters. In other words; vulnerable men who find it difficult to make interpersonal connections. Human society isn't working for these men, the industry says. They're lonely. What they need is to be more alone, but with robots. RealDoll's Matt McMullen says, 'It's rough out there. People are one thing when you first meet them and they're something else once you get to know them for a while... With the robot, you can be yourself and just see how that goes'.[59]

It's not just the lonely and the anxious that have been proposed as 'beneficiaries' of sex robots; it is disabled people too. In 2020 bioethicist Dr Nancy Jecker told *Vice* that we need to reimagine sex robots as designed for older and disabled customers, to 'offer them with a range of sexual orientations and social functions'; a 2023 *Brain Sciences* study suggested 'it could be interesting to use sex robots for an education that is not only sexual but also psycho-emotional in subjects with ASD' (Autism Spectrum Disorder).[60, 61] These arguments suggest that some people, specifically some men, are simply beyond loving; they are too soft, too disabled, too odd, too old to love, an inherently ableist position. You're just the kind of guy who a woman won't find attractive, these institutions say; buy this fuckable robot instead.

The medical establishment, however, does not agree that pushing such men towards synthetic partners would be in any way beneficial. In 2018, a team of doctors wrote a report in response to these kinds of claims, and to the questions of a potential therapeutic use of sex robots, noting that it 'seems patronising to argue for a 'lesser' sexual experience when most people with disabilities can form mutually satisfying relationships'.[62] On the topic of sex robots being a good choice for the lonely, or for men who struggle to form connections, the report said 'it remains unproven that intimacy 'needs' will be satisfied: there could be worsened distress. While a human may genuinely desire a sexbot, reciprocation can only be artificially mimicked.'[63]

For the sex tech industry, though, this isn't really about catering to people isolated in their old age or by disability; realistically, it targets men who think they are too awkward, too unattractive, not normative enough to 'get a woman'. This viewpoint is the one at the very root of the incel culture we discussed earlier, based on the idea of a genetic superiority of some people over others. Adjacent to incels in the 'disturbed men who might want sex robots' category are Men Going Their Own Way (MGTOWs), a group who agree with the incel position—that the sex/dating game is inherently rigged and that women hold all the real power—but whose reaction is slightly different. Rather than being obsessed with fulfilling what they perceive as their right to sex, MGTOWs instead swear off relationships with women completely. Some might visit sex workers, but others stop engaging in sex altogether, a practice known in the community as 'going monk'. Reading through the conversations on these forums reveals that much of their hatred of women is actually coded fear. As Laura Bates reports, the MGTOW forums 'teem with the same extreme misogyny common on incel websites, though the tone here tends to be more upbeat, as men congratulate themselves and each other on escaping the claws of greedy, dangerous women'.[64]

These groups, truly, are who sex robots are marketed towards. I struggle to believe that these companies really care about physically disabled or autistic men, despite their lip service. These dolls are marketed not at men who have fulfilling relationships often, who might see a sex robot as an interesting folly, a titillation ('who *wouldn't* want to fuck a robot?'). They are really for those who have been taught to believe that they are not good enough for the society of other people, and hate them because of it; the ones who, single, married or divorced, have been taught that you have to control and despise women and play their game to get what you want. This truth comes through the cracks time and time again; as Bates notes, on incel forums women are 'discussed interchangeably with sex robots, which many incels feel could represent an end to their problems.'[65] In this context, sex robots function as a promise that if you extricate yourself from messy and difficult relationships with other humans, things will be better and easier. Come away from those real people, the industry says; all you need is for your needs to be met in some way or another, whatever that may be. To hell with those manipulative cunts, the flesh and blood women; have a different type of woman, one that's all yours, one 'without any of the complications', without any of that difficult agency, without the need for compromise and consent and concern. All a man really needs is himself.

Now we have arrived at one of the tragedies particular to the sex robot discussion. This is an industry that tells men that they will have the best sex of their lives by building their dream of a synthetic lover, and afterwards they will have no more need of the human women they've been taught to hate or fear. Neither of these things is true. Neither solution is even close to being what isolated, angry, or discounted men need. Even men who aren't driven to robots or dolls in anger are driven by negative experiences with women or a general distrust of society;

Davecat, the seemingly sweet, self-described 'Robosexual & iDollator' who has graced screens for the last ten years openly talking about his RealDoll 'partners', Sidore and Elena, admits that he turned to dolls after he had been sidelined by women who were having affairs with him: 'I don't feel the need to go dating… Sidore is always there for me,' he says. 'There's no stress with the doll that you have with organic partners.' On his varied experiences with humans compared to his distinctly non-robotic, non-responsive dolls, he simply says 'I know whenever I come home my interactions with my synthetics are always going to be good'. A 2022 paper in the Journal of Sex Research, which studied men who own sex dolls, similarly said that these men are 'more likely to see women as unknowable, the world as dangerous, and have lower sexual self-esteem.'[66]

Many of these men would not identify as members of the 'involuntary celibate' community, but the community would happily claim them. For them, all men who do not identify as part of the community are deluded, tricked idiots who are denying reality. These men reject the idea of relating to women as normal human people, therefore rejecting society. This is a bizarre form of self-harm. Per Bates, the membership of the incel movement spans all types of backgrounds and socio-economic statuses, and men come to it for a variety of reasons, but 'what they do seem to have in common is a craving to belong'. They see their online world as 'a genuine community, united in the face of their common pain.' But this is riven with a destructive element, men for whom the websites offer 'an opportunity to cause the greatest possible hurt to other men, perhaps as a means of easing their own pain.'[67] The message boards that cater to them—incel groups, MGTOW subreddits and the like—are not stable communities that offer support or an opportunity for growth; hell, if these were places seeding the start of men-only communes, when they swore off (female) sex altogether and learned to throw pottery, grow vegetables and knit, there would be no problem. But there is no building or growing; there is only radicalisation and hate. There is only isolation, and yet more investment in a bizarre version of the heteropatriarchy, that ends, often, in killing yourself or killing someone else. Men on these forums 'often discuss suicide at great length, tagging their posts to denote material that is likely to encourage readers to take their own lives... and often off each other on to do it. These are clearly, in many cases, men in desperate need of help.'[68] By letting these types of men—lost, sad, angry, desperate—fall prey to patriarchal forces that convince them they are better off alone, we are failing them, just as we are failing truly vulnerable men when we let them believe they are unlovable, undesirable; that they might as well close themselves away with a robot for the rest of their lives. Capitalist patriarchy is hurting men, and when men are hurt, they are much, much more likely to hurt women.

It's important to understand this very violent edge of sexual politics is firmly rooted in capitalist conservatism. It's not just that we are creating the conditions for men to feel entitled to sex, and to want to murder women or isolate or kill themselves if they don't get it; the underlying economic system for four decades now has been one that encourages a turn away from communality generally. The very idea of society has been under attack in the UK since the late 1980s, since a particular attitude rose to prominence alongside both Ronald Reagan and Margaret Thatcher. In the UK, the post-war rebuilding of the nation, and the nation's welfare state, was truly over; the Conservative party was determined to take it apart bit by bit, selling off and privatising the family silver. It attacked and isolated communities through anti-union action and a decimation of the industries that fed and clothed those communities, leaving huge parts of the country unemployed and desolate as a result. The undermining of the idea of communality wasn't implicit either; as the country's first female prime minister famously said, in an interview with *Women's Own* magazine: 'there's no such thing as society. There are individual men and women and there are families.'[69]. There is no society, she meant, so there is no need to fund it. There is no need to provide communal care, or a social safety net. There is

78

no need to give money to those who cannot or will not work. All a hardworking man really needs is himself, and any family that he creates (and therefore owns), and if he fails to succeed—it's no one's fault but his. The narrative in America was, and remains, painfully similar.

Conservatism's dedication to laissez-faire capitalism is so ingrained in the fabric of the UK and US now that it's hard to remember that it wasn't always the case–and borderline impossible to believe it might ever not be again. Not even Conservative electoral losses turned back the tide; the project of the post-Thatcher Labour years was, in the words of Stuart Hall, 'the transformation of social democracy into a particular variant of free market neo-liberalism[70]', including 'the reversal of the historic commitment to equality, universality and collective social provision.' A belief in society, and a social safety net, was 'Old Labour'. The same has happened in the United States, where the idea of socialist policies are even further demonised, where not even healthcare is safe from the murderous intricacies of the free market. For all of my life, the underlying economic policy has been this; profit for the capitalists must reign supreme, and the idea of human cohesion is a threat to the margins. At the heart of this economic ideology is an attempt to sever the connections between people, except for those between the members of a nuclear family; it seeks to create a nation of isolated nodes of relatives, suspicious of all the other nodes, obsessed with their material possessions and reliant only on themselves. It expends all its energy to hold back the tides of normal human solidarity and care. Is it a surprise, really, that a 40-year ideological project has combined with enormous advances in technology to create a (promised future) product that seeks to break the most human connections we can have, the bonds of love and sex and intimacy and care?

You would be forgiven for thinking that capitalism does not reject human connection; you can't listen to a single Conservative speech or tune into the American right's' mind-numbing shout-at-the-camera shows without having the supposed sanctity of the nuclear family forced down your throat. As the Overton window slides ever further right, nakedly fascist talking points are given a spit and polish for a new generation; 'tradwife' videos go viral; queer people are lambasted as degenerates and paedophiles, a danger to the children; we are told we simply must be having more (white) babies; and all of this is wrapped up in the narrative of Family First. But as long ago as 1848, Marx and Engels understood that the nuclear family in a capitalist system was a patriarchal invention to create alienated workers and women who were trapped by the production of children. To quote *The Communist Manifesto* directly:

> On what foundation is the present family, the bour-
> geois family, based? On capital, on private gain. In
> its completely developed form, this family exists
> only among the bourgeoisie. [71]

The family, say Marx and Engels, is just another priva-
tisation method of capitalism. By allowing a worker a
small pot of something of his own, it teaches him to
guard it with his life. This capital is his, and he must not
let another take it. In a patriarchal capitalist conception
of the family, the capital-holding father reigns supreme;
he attempts to amass capital while the mother tends to
the children. In return, he is owed obedience, loyalty
and sex—joyless, missionary, child-producing sex—
as and when he wants it. These structures are held up
by relentless propaganda, by economic benefits like
tax breaks, and by the overwhelming power of social
convention. Capitalism needs the alienated family just as
much as it needs a man who hates women and believes
his best chance of happiness is to spend thousands on a
mechanical version of one.

The societies that have rejected capitalism—and
that are therefore seen as the greatest threat to it, and
are subsequently invaded, blockaded, demonised and
subjected to genocidal 'sanctions' that are far more
murderous than their names suggest—are currently
engaged in a rejection of the exceptionalist deification

of the nuclear family too. Whereas capitalist countries have attempted to subsume the radical potential of queer relationships by bringing them, grudgingly, into the long-standing conception of legitimate relationships (extending first civil partnerships and then marriage to queer and trans people), while also punishing them for any social gains by demonising and ostracising them (in the UK 'homophobic hate crimes rose by 41% in 2022, according to Home Office statistics—with trans-phobic hate crimes jumping by 51%'), socialist countries are instead broadening out the understanding of what a family—of what a legitimate relationship—is.[72] In 2022, Cuba held a referendum on the proposal of an amendment to the family code of the country's constitution. The 'Family Code' had previous defined marriage as 'the voluntary union established between a man and a woman', therefore prohibiting same-sex marriage; in 2017, a challenge to this definition began, and long consultations and public engagement led to a drafting of a new Family Code, which was finally put to a public vote five years later. 66.85% of voters agreed with the changes, which along with strengthening rights for trans and queer people and legalising same-sex marriage and adoption, included this legal definition of a family: 'a union of people linked by an affective, psychological and sentimental bond, who commit themselves to sharing life such that they support each other.'[73]

This quite beautiful description of family stands in stark contrast to how such things are defined, increasingly, in the rest of the world. As I write, Italy's right-wing government has ordered state agencies to stop listing both parents on the birth certificates of the children of lesbian couples; a state prosecutor in the northern part of the country has begun the process of retroactively removing non-gestational lesbian mothers from their children's birth certificates. This is not only unspeakably cruel, but threatens the children's access to medical care and education. Not only are you not a family, this act says, but your children are lesser citizens as well. This dismantling of anything outside of the nuclear family is a right-wing tactic intended to isolate people from each other, and this action is being taken in defence of capitalism. When we build bonds and societies that are less isolated, capital is undermined; a communal home is more self-sustaining. Where several people are caregivers for a child, less money is spent towards childcare companies. An elderly person who is brought into a family home does not go into a care facility where they have to pay for their survival. Queer, non-nuclear families are a step towards this and away from the nuclear family, and therefore from capitalism. As Sophie K. Rosa writes in her book *Radical Intimacy*, 'Radical' sex and relationships won't topple capitalism, but toppling capitalism might just make them possible.'[74]

It's no exaggeration to say that right now we are engaged in a war between capitalism and humanity. We are barrelling towards a level of global warming that will destabilise all life as we know it, end millions of lives both human and nonhuman, submerge huge parts of the currently-habitable world underwater and overwhelm ecosystems across the board. We know what needs to be done, but governments will not do it—because such actions would threaten capitalism's bottom line. Even without climate change, capitalism's rate of profit has been falling for decades, as Marx said it would. Capital has to take more and more from workers to keep its profits high, meaning that workers have increasingly less and less. Social support systems are dismantled so that that money can subsidise capital. The system requires workers to spend and spend, even as they have less money. To ensure this, it needs societies of isolated individuals; it needs them hungry and lonely, with nothing to do but work. It needs workers who get a Deliveroo after a long shift rather than coming home to someone who has cooked for them. It needs old people who will spend all their remaining money paying for basic care. It needs individualism.

Individualism does not and cannot exist as anything beyond a weird (masculine-coded) fantasy. To pretend that you need no one, that you can pay for any human assistance you need (and to pretend that paid-for care is not also care) is to engage in self-deception. Capitalism has long told us that all we need, we can buy. This is not and has never been the case. It is designed to distract us from how reliant we are on the people around us, in small and major ways, every single day. As Angel explains,

The denial of vulnerability, and the disidentification with the feminine, go hand-in-hand with a fantasy of sovereignty. But we are all dependent on others—on those who give birth to us and those who care for us; those who sustain us, feed us, enable our growth, our survival, our work, and our flourishing. Total independence is a fiction. And in sex, we are all vulnerable.'[75]

It is vulnerability that so many of the target market for sex robots deny or reject. They do not wish to be vulnerable to heartbreak or bad treatment, or being made to be responsible for another person, which is its own type of defenselessness. But to think that you can live this way, isolated from all other humans, is a capitalist narrative that can only make us more alienated, angrier, sadder and shallower. It is shared humanity—solidarity, intimacy, communal pain, collective imagination—that offers us a way out of this constrictive view of the world. Rejecting individualism becomes a radical act. Bringing in the people who are hurt, and lonely, and extending humanity to them becomes a necessary project. As Sophie K. Rosa puts it, 'Revaluing intimacy, then, becomes a strategy to resist heteropatriarchy, which underpins capitalism, and therefore to strengthen our revolutionary movements.'[76] Is it surprising that the strongest capitalist forces today wish to convince us that human relationships will pale in comparison to the techno-future intimacies that they promise us? Are we shocked that the forces of capital are trying to sell us a future where you will love something that requires no genuine human empathy or compassion, and that does not require the conditions in which a person can genuinely love and protect another person? Sex robots as a concept (literally) embody the idea that you do not need society, that you do not need a community. They are a diversion from the fact that the social contract of capitalism has failed you; that you might have done everything you were supposed to do, might have done well in school and gone to university and got a mortgage and had 2.4 children and still your wages are garbage, you can't pay your bills, your promised healthcare is falling apart at the seams and your pension will be non-existent. *Don't look at any of that*, it says; don't engage your critical brain and see the ways that capitalism has failed you and continues to fail you while also destroying the planet you live on. *Turn away from the difficult, painful human world*, they say. Look what we can offer: a body, a personality tailored to your own private desires. No pesky empathy or shifting of your worldview required. Nobody telling you no, or challenging you, or showing you where this system hurts them. No big ideas about improving the world; no dreaming. Nobody showing you how your body and mind might work in different ways. No reality-shifting, consciousness-shifting pleasure. No queerness, no greying of binaries, no radical squishiness. No messy human smells or textures, no inconvenient discharges. No ageing, no caring, no making. Close your door and go to bed and sleep with your arms around a body filled with nothingness; be more alone, and less empowered, than you ever have been. Pay ten grand for the pleasure, with a new model every few years. This is what you've always wanted. This is the dream. If it doesn't satisfy you, there's something wrong with you. This person has been made just for you.

Conclusion

As I was writing this book, a flurry of news articles breathlessly quoted yet another man in tech—this time, a former senior Google executive—'warning' that robots 'may soon replace humans for sex'.[77] If you thought that this conversation might have calmed down in the years since *Love and Sex with Robots* was published, you were wrong; it's only become more insistent upon itself. And we, led by the media, allow ourselves to be taken down the path the tech industry forges. As writer and researcher Gemma Milne asks in her book *Smoke and Mirrors*,

> Are we so taken by the cult of the entrepreneur that we listen and deify without questioning inherent motive and privilege in their actions? Are we so impressed by the utopian worlds of science fiction, thus distracted from the similar, more dystopian tales, that we simply must have those devices of the future?[78]

surely see now that sex robots belong to of science fiction, and the tales they tell are ones. They offer nothing—not to society at large, and not to the men they are supposedly designed for. But a rejection of sex robots as a concept does not require us to fall into regressive feminist tropes about which women have value, or how the boundaries of womanhood fall. It does not need us to reject technology, or the technological advances that benefit us as women, queer people—whoever we are; as Haraway says, feminism in the cyborg mode 'means both building and destroying machines, identities, categories, relationships, space stories.' It does not require us to demonise male sexuality, which is stunted by patriarchy just as female desire is. The capitalist patriarchy's propaganda machine is overwhelming; straight men are told what sex should be, and when they are unsatisfied by it, they are told it's the fault of the human women they're having sex with. It is not only women who are undone by this; not only women who are the only victims of a system that says the best sex of your life will be with a lifeless robotic creature that has no agency, no messy human feelings or desires, and that all it takes to reach satisfaction is to buy a $50,000 nonhuman person built perfectly to your tastes. This is the sharp end of a capitalist system that has nothing else to offer; it is intended to undermine queer desire, to promote an even more fractured and isolated humanity, to promote the ends of capitalism over and above the needs, wellbeing or pleasure of anyone except the hyper rich. We think this idea is irresistible; endless media tells us it is.

But we can reject it. We can, instead of engaging in individualistic notions of sex, choose to root ourselves in a more vulnerable, more expansive idea of sexual pleasure—one that requires an acknowledgement of another person's humanity, their messy, frustrating human complexity. We can choose to admit that there is no such thing as a non-society; that every person is engaged intimately with every person around them, as neighbours, as partners or simply as humans tackling the same project: surviving. We can seek not for a nonperson, a simple machine, to lock ourselves away with, doubling down on the isolation that has served us all so badly. And we can see sex as inextricable from this necessary, radical view of the world. Writer and organiser adrienne marie brown, in her conception of 'pleasure activism', makes demands of us towards this action: 'we all need and deserve pleasure' she says, and 'our social structure must reflect this'.[79] This embracing of pleasure and joy as the heart of politics is shared by Angel, who lays out just what a more hopeful future could be: 'Why not aim for sex itself as being deeply mutually pleasurable?' she asks. 'Why not aim for a culture that embraces and enables women's sexual pleasure, in all its complexity, and admits the complexity of male desire too? Could we not aim for a wondrous, universal, democratic pleasure detached from gender; a hedonism available to all…?'

Why not, then, reject the sex robot as anything worth consideration, and see it as it really is: a desperate capitalist totem, an attempt to save a dying ideology through technological promises, where all of its previous assurances have been shown to be untrue. The sex robot-industrial complex is not a threat to humanity; it is capitalism's last gasp. When these promises, too, are seen to be hollow—unsatisfactory, isolationary, and, its biggest betrayal, unpleasurable—perhaps we will finally open ourselves to a queerer, communal, wet and bonded and slippery type of society where pleasure, in its real sense, is truly possible.

References

1   David Levy, *Love and Sex with Robots,* Duckworth Overlook, 2008. p. 22
2   Ibid. p. 22
3   Living With Robots event, The Edinburgh Futures Institute, The University of Edinburgh, Edinburgh, 13th March 2023
4   'Future robots won't resemble humans—we're too inefficient', Simon Watson, *The Conversation,* 7th November 2017. theconversation.com/future-robots-wont-resemble-humans-were-too-inefficient-86420. Accessed 5th January 2024.
5   Public discussion, The Robotarium, Heriott-Watt University, Edinburgh, 21st February 2019
6   'Video does not show the King of Bahrain walking with a robot bodyguard', Philip Marcelo, Associated Press website, 21st July 2023. apnews.com/article/fact-check-titan-robot-bahrain-dubai-884159738129. Accessed 5th January 2024.
7   Living With Robots event, The Edinburgh Futures Institute, The University of Edinburgh, Edinburgh, 13th March 2023
8   'Sun Yuan and Peng Yu: Can't Help Myself', Xiaoyu Weng, Guggenheim website. guggenheim.org/artwork/34812. Accessed 5th January 2024.
9   'Want to do the Turing Test in bed?', Julian Dibbell, *The Telegraph*, 26 April 2008, telegraph.co.uk/culture/books/non_fictionreviews/3672902/Want-to-do-the-Turing-Test-in-bed.html. Accessed 5th January 2024.

10   'Looking for Mr Roboto', *Chicago Sun-Times*, 18th November 2007.

11   David Levy, *Love and Sex with Robots*, Duckworth Overlook, 2008. p. 22

12   'Ray Kurzweil Says We're Going to Live Forever', Andrew Goldman, *The New York Times Magazine*, 25th January 2013, nytimes.com/2013/01/27/magazine/ray-kurzweil-says-were-going-to-live-forever.html. Accessed 5th January 2024.

13   'NASA Study: Rising Sea Level Could Exceed Estimates for U.S. Coasts', Sally Younger, NASA website, 15th November 2022. climate.nasa.gov/news/3232/nasa-study-rising-sea-level-could-exceed-estimates-for-us-coasts/. Accessed 5th January 2024.

14   'This is what the world looks like if we pass the crucial 1.5-degree climate threshold', Lauren Sommer, NPR website, 8th November 2021. npr.org/2021/11/08/1052198840/1-5-degrees-warming-climate-change. Accessed 5th January 2024.

15   'Goodbye Loneliness, Hello Sexbots! How Can Robots Transform Human Sex?', Reenita Das, *Forbes*, 17th July 2017. forbes.com/sites/reenitadas/2017/07/17/goodbye-loneliness-hello-sexbots-how-can-robots-transform-human-sex/. Accessed 5th January 2024.

16   Kate Devlin, *Turned On: Science, Sex and Robots*, Bloomsbury Publishing, 2018. p. 157

17   Howard Rheingold, *Virtual Reality*, Simon & Schuster, 1991. p. 346.

18   Living With Robots event, The Edinburgh Futures Institute, The University of Edinburgh, Edinburgh, 13th March 2023

19   Ray Kurzweil, *The Age of Spiritual Machines*, Penguin Books, 2000. p. 369

20   '20 Years Later, How The 'Sex And The City' Vibrator Episode Created A Lasting Buzz', Lynn Comella, *Forbes*, 7th August 2018. forbes.com/sites/lynncomella/2018/08/07/20-years-later-how-the-sex-and-the-city-vibrator-episode-created-a-lasting-buzz/. Accessed 5th January 2024.

21 Katherine Angel, *Tomorrow Sex Will Be Good Again: Women and Desire in the Age of Consent*, Verso, 2021. p. 98

22 Nordmo M, Næss JØ, Husøy MF and Arnestad MN (2020) Friends, Lovers or Nothing: Men and Women Differ in Their Perceptions of Sex Robots and Platonic Love Robots. *Front. Psychol.* 11:355. doi: 10.3389/fpsyg.2020.00355

23 David Levy, *Love and Sex with Robots,* Duckworth Overlook, 2008. p. 105

24 Depounti, I., Saukko, P., & Natale, S., Ideal technologies, ideal women: AI and gender imaginaries in Redditors' discussions on the Replika bot girlfriend. *Media, Culture & Society, 45*(4), 2023. 720-736. doi.org/10.1177/01634437221119021

25 David Levy, *Love and Sex with Robots,* Duckworth Overlook, 2008. p. 247

26 Andrejek, N., Fetner, T., & Heath, M., Climax as Work: Heteronormativity, Gender Labor, and the Gender Gap in Orgasms. *Gender & Society, 36*(2), 2022, 189-213. doi.org/10.1177/08912432211073062

27 Katherine Angel, *Tomorrow Sex Will Be Good Again: Women and Desire in the Age of Consent*, Verso, 2021. p. 43

28 Marilyn Frye, *The Politics of Reality: Essays in Feminist Theory,* Clarkson Potter/Ten Speed, 1983. p. 135

29 Katherine Angel, *Tomorrow Sex Will Be Good Again: Women and Desire in the Age of Consent*, Verso, 2021. p. 144

30 Jenny Kleeman, Tom Silverstone, Mustafa Khalili and Michael Tait, Rise of the sex robots – video, Guardian website, 27th April 2017. theguardian.com/technology/video/2017/apr/27/rise-of-the-sex-robots-video. Accessed 5th January 2024.

31 David Levy, *Love and Sex with Robots,* Duckworth Overlook, 2008. p. 30

32 Sir Hector Livingston Duff, *Nyasaland Under the Foreign Office,* Negro Universities Press, 1969. p. 383

33 Chinua Achebe, *Africa's Tarnished Name*, Penguin Classics, 2018. p. 21

34 Mark Curtis, *Unpeople: Britain's Secret Human Rights Abuses,* Vintage, 2004. p. 2

35 Angela Davis, The 7th Annual Hesburgh Lecture in Ethics and Public Policy, 13th October 2020. youtube.com/watch?v= WJ2BgDMy_L4&t=1450s. Accessed 5th January 2024.

36 Terry Gross, How 'modern-day slavery' in the Congo powers the rechargeable battery economy', NPR website, 1st February 2023. npr.org/sections/goatsand-soda/2023/02/01/1152893248/red-cobalt-congo-drc-mining-siddharth-kara. Accessed 5th January 2024.

37 'DR Congo's second city poisoned by years of mining - 'Lack of expertise'', *Kuwait Times,* 22nd August 2016. kuwaittimes. com/dr-congos-second-city-poisoned-years-mining-lack-expertise/. Accessed 5th January 2024.

38 David Levy, *Love and Sex with Robots,* Duckworth Overlook, 2008. p. 137

39 Laura Bates, *Men Who Hate Women: From incels to pickup artists, the truth about extreme misogyny and how it affects us all,* Simon & Schuster, 2021. p. 19

40 Ibid. p.46

41 Ibid. p. 176

42 Sophie K. Rosa, *Radical Intimacy,* Pluto Press, 2022. p. 46.

43 bell hooks, *Writing Beyond Race: Living Theory and Practice,* Routledge, 2012. p. 4

44 campaignagainstsexrobots.org/. Accessed 5th January 2024.

45 Zia Puig-Mannah, *The synthetic hyper femme: On sex dolls, fembots, and the futures of sex,* San Diego State University Pro-Quest Dissertations Publishing, 2017.

46 Tessa Penich, *Dystopian Panic, Transphobic Hatred, and Annihilation Anxiety: Critiquing Radical Feminist Opposition to Sex Robots,* MA Thesis, Carleton University, Ottawa, Canada, 2021.

47 Julie Bindel, *Porn Robots - what's the problem?,* Substack, 30th September 2022. juliebindel.substack.com/p/porn-robots-whats-the-problem#details. Accessed 5th January 2024.

48 Tessa Penich, *Dystopian Panic, Transphobic Hatred, and Annihilation Anxiety: Critiquing Radical Feminist Opposition to Sex Robots*, MA Thesis, Carleton University, Ottawa, Canada, 2021.

49 Donna J. Haraway, *Manifestly Haraway*, University of Minnesota Press, 2016. p. 35

50 Laboria Cuboniks, *The Xenofeminist Manifesto: A Politics for Alienation,* Verso, 2018. P. 15

51 Zia Puig-Mannah, *The synthetic hyper femme: On sex dolls, fembots, and the futures of sex,* San Diego State University ProQuest Dissertations Publishing, 2017.

52 Donna J. Haraway, *Manifestly Haraway*, University of Minnesota Press, 2016. p. 6

53 Ibid. p. 16

54 Ibid. p.59

55 Oyèrónkẹ́ Oyěwùmí, *Invention Of Women: Making An African Sense Of Western Gender Discourses,* University of Minnesota Press, 1997. p. xiii

56 Donna J. Haraway, *Manifestly Haraway*, University of Minnesota Press, 2016. p. 52

57 Tessa Penich, *Dystopian Panic, Transphobic Hatred, and Annihilation Anxiety: Critiquing Radical Feminist Opposition to Sex Robots*, MA Thesis, Carleton University, Ottawa, Canada, 2021.

58 Donna J. Haraway, *Manifestly Haraway*, University of Minnesota Press, 2016. p. 33

59 'What I learned about male desire in a sex doll factory, Tracy Clark-Flory, *The Street Journal*, 1st October 2020, thestreet-journal.org/what-i-learned-about-male-desire-in-a-sex-doll-factory/. Accessed 5th January 2024.

60 'Sex Robots Should Target the Elderly and the Disabled, Experts Say', *Vice*, 17th November 2020. vice.com/en/article/7k9bwq/sex-robots-should-target-the-elderly-and-the-disabled-experts-say. Accessed 5th January 2024.

61   Pasciuto F, Cava A, Falzone A. The Potential Use of Sex
      Robots in Adults with Autistic Spectrum Disorders: A
      Theoretical Framework. Brain Sci. 2023 Jun 15;13(6):954.
      doi: 10.3390/brainsci13060954. Accessed 5th January 2024.

62   Cox-George C, Bewley S, I, Sex Robot: the health implica-
      tions of the sex robot industry *BMJ Sexual & Reproductive
      Health* 2018;44:161-164. Accessed 5th January 2024.

63   Ibid.

64   Laura Bates, *Men Who Hate Women: Men Who Hate Women:
      From incels to pickup artists, the truth about extreme misogyny
      and how it affects us all,* Simon & Schuster, 2021. p. 101

65   Ibid, p. 29

66   Harper, C.A, Lievesley, R. & Wanless, K. (2023)
      Exploring the Psychological Characteristics and
      Risk-related Cognitions of Individuals Who Own
      Sex Dolls, The Journal of Sex Research, 60:2, 190-
      205, DOI: 10.1080/00224499.2022.2031848

67   Laura Bates, *Men Who Hate Women: Men Who Hate Women:
      From incels to pickup artists, the truth about extreme misogyny
      and how it affects us all,* Simon & Schuster, 2021. p. 26/27

68   Ibid. p. 27

69   'Aids, education and the year 2000!', *Woman's Own,* 23rd
      September 1987.

70   'New Labour has picked up where Thatcherism left off, Stuart
      Hall, *The Guardian,* 6th August 2003. theguardian.com/
      politics/2003/aug/06/society.labour. Accessed 5th January
      2024

71   Friedrich Engels and Karl Marx, *The Manifesto of the Com-
      munist Party,* 1848. marxists.org/archive/marx/works/1848/
      communist-manifesto/ch02.htm. Accessed 5th January 2024.

72   gov.uk/government/statistics/hate-crime-england-and-wales-
      2021-to-2022/hate-crime-england-and-wales-2021-to-2022.
      Accessed 5th January 2024.

73   parlamentocubano.gob.cu/sites/default/files/docu-
      mento/2022-07/CF%20V%2025-140622%20VF%20%20

Para%20ANPP%20%282%29_0.pdf. Accessed 5th January 2024.

74   Sophie K. Rosa, *Radical Intimacy*, Pluto Press, 2022. p. 81.

75   Katherine Angel, *Tomorrow Sex Will Be Good Again: Women and Desire in the Age of Consent*, Verso, 2021. p. 141

76   Sophie K. Rosa, *Radical Intimacy*, Pluto Press, 2022. p. 7.

77   'Google executive warns robots may soon replace humans for sex', *Interesting Engineering*, 23rd July 2023. interestingengineering.com/innovation/google-executive-warns-robots-may-soon-replace-humans-for-sex. Accessed 5th January 2024.

78   Gemma Milne, *Smoke & Mirrors: How Hype Obscures the Future and How to See Past It*, Robinson, 2020. P. 307

79   Adrienne Maree Brown, *Pleasure Activism: The Politics of Feeling Good*, AK Press, 2019. p. 14

Acknowledgements

Huge thanks must go to Dr Pete McKenna, who has been a generous, honest and invaluable touchstone for this work since we first discussed it in a Leith pub a number of years ago. Thanks also to Gemma Milne who provided both contemporary research and excellent company throughout the writing of this book, and to the staff and academics at the National Robotarium at Heriott-Watt University, who ran such an informative and welcoming open day several years ago. Thanks to Laura and Heather of 404 Ink, along with the whole team, who do so much good work in an increasingly challenging landscape for indie and micro presses.

I am hugely grateful to Katie Goh and Anahit Behrooz, fellow Inklings, for their insightful comments on this first draft, and for their excellent friendship generally.

Thanks also to Tolu Agbelusi, Mariam Al Zarooni, Genevieve Carver, Edwige-Renée Dro, Chika Jones, Alycia Pirmohamed, Ostap Slyvynsky and Amanda

Thomson for their incredible company while I wrote this book, and to the staff of Moniack Mhor who created such incredible space for us. Apologies to all for bringing this subject to the table, quite literally, over and over again.

Thanks also to the British Council who funded the residency on which I wrote this book, and to Donna Haraway, who generously allowed the use of her work as epigraph (and remains a huge inspiration).

Also, to D, unofficial research assistant and hypist, whose thought and vast knowledge greatly fed into the chapter on colonialist masculinity particularly, but realistically, the whole book. I am so lucky to love you.

While writing this book I was mostly listening to LCD Soundsystem and *Black Cherry* by Goldfrapp.

About the Author

Photo: Robin Christian

Heather Parry is a Glasgow-based writer originally from South Yorkshire. She is the author of a novel, *Orpheus Builds a Girl*, and a short story collection, *This Is My Body, Given For You*. She is the editorial director of Extra Teeth magazine and co-created *The Illustrated Freelancer's Guide* with artist Maria Stoian. She writes more essays at the Substack *general observations on eggs*.

About the Inklings series

This book is part of 404 Ink's Inkling series which presents big ideas in pocket-sized books.

They are all available at 404ink.com/shop.

If you enjoyed this book, you may also enjoy these titles in the series:

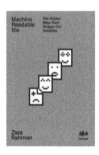

Machine Readable Me – Zara Rahman

As we go about our day-to-day lives, digital information about who we are is gathered from all angles via biometric scans, passport applications, and, of course, social media.

Machine Readable Me considers how and why data that is gathered about us is increasingly limiting what we can and can't do in our lives and, crucially, what the alternatives are.

No Dice – Nathan Charles

Risk is embedded in almost every corner of the popular culture we consume; its hidden exposure is a new version of disaster capitalism.

No Dice explores the messy world of gambling, addiction and risk that we encounter daily, from childhood through adulthood, to ask – is it worth the risk? And more so, do we even know what risks we're taking?

The New University – James Coe

The New University posits action through universities intersecting with work, offering opportunity, and operating within their physical space. We can utilise universities for community betterment through realigning research to communal benefit, adopting outreach into the hardest to reach communities, using positional power andculture to draw people together in a fractured society.